T0142082

Studies in Computational Intelligence

Volume 574

Series editor

Janusz Kacprzyk, Polish Academy of Sciences, Warsaw, Poland
e-mail: kacprzyk@ibspan.waw.pl

About this Series

The series "Studies in Computational Intelligence" (SCI) publishes new developments and advances in the various areas of computational intelligence—quickly and with a high quality. The intent is to cover the theory, applications, and design methods of computational intelligence, as embedded in the fields of engineering, computer science, physics and life sciences, as well as the methodologies behind them. The series contains monographs, lecture notes and edited volumes in computational intelligence spanning the areas of neural networks, connectionist systems, genetic algorithms, evolutionary computation, artificial intelligence, cellular automata, self-organizing systems, soft computing, fuzzy systems, and hybrid intelligent systems. Of particular value to both the contributors and the readership are the short publication timeframe and the world-wide distribution, which enable both wide and rapid dissemination of research output.

More information about this series at http://www.springer.com/series/7092

Oscar Castillo · Patricia Melin
Editors

Fuzzy Logic Augmentation of Nature-Inspired Optimization Metaheuristics

Theory and Applications

 Springer

Editors
Oscar Castillo
Patricia Melin
Division of Graduate Studies and Research
Tijuana Institute of Technology
Tijuana
Baja California
Mexico

ISSN 1860-949X ISSN 1860-9503 (electronic)
ISBN 978-3-319-38546-4 ISBN 978-3-319-10960-2 (eBook)
DOI 10.1007/978-3-319-10960-2

Springer Cham Heidelberg New York Dordrecht London

© Springer International Publishing Switzerland 2015
Softcover reprint of the hardcover 1st edition 2015
This work is subject to copyright. All rights are reserved by the Publisher, whether the whole or part of the material is concerned, specifically the rights of translation, reprinting, reuse of illustrations, recitation, broadcasting, reproduction on microfilms or in any other physical way, and transmission or information storage and retrieval, electronic adaptation, computer software, or by similar or dissimilar methodology now known or hereafter developed. Exempted from this legal reservation are brief excerpts in connection with reviews or scholarly analysis or material supplied specifically for the purpose of being entered and executed on a computer system, for exclusive use by the purchaser of the work. Duplication of this publication or parts thereof is permitted only under the provisions of the Copyright Law of the Publisher's location, in its current version, and permission for use must always be obtained from Springer. Permissions for use may be obtained through RightsLink at the Copyright Clearance Center. Violations are liable to prosecution under the respective Copyright Law.
The use of general descriptive names, registered names, trademarks, service marks, etc. in this publication does not imply, even in the absence of a specific statement, that such names are exempt from the relevant protective laws and regulations and therefore free for general use.
While the advice and information in this book are believed to be true and accurate at the date of publication, neither the authors nor the editors nor the publisher can accept any legal responsibility for any errors or omissions that may be made. The publisher makes no warranty, express or implied, with respect to the material contained herein.

Printed on acid-free paper

Springer is part of Springer Science+Business Media (www.springer.com)

Preface

We describe in this book, recent advances on fuzzy logic augmentation of nature-inspired optimization metaheuristics and their application in areas, such as intelligent control and robotics, pattern recognition, time series prediction, and optimization of complex problems. The book is organized into two main parts, which contain a group of papers around a similar subject. Part I consists of papers with the main theme of theoretical aspects of fuzzy logic augmentation of nature-inspired optimization metaheuristics, which basically consists of papers that propose new optimization algorithms enhanced using fuzzy systems. Part II contains papers with the main theme of application of optimization algorithms, which are basically papers using nature-inspired techniques to achieve optimization of complex optimization problems in diverse areas of application.

In the part of theoretical aspects of fuzzy logic augmentation of nature-inspired optimization metaheuristics, there are seven chapters that describe different contributions that propose new models and concepts, which can be the considered as the basis for enhancing nature-inspired algorithms with fuzzy logic. The aim of using fuzzy logic is to provide dynamic adaptation capabilities to the optimization algorithms, and this is illustrated with the cases of the bat algorithm, cuckoo search, and other methods. In the part of applications of fuzzy nature-inspired algorithms there are five chapters that describe different contributions on the application of the nature-inspired algorithms to solve complex optimization problems. The nature-inspired methods include variations of ant colony optimization, particle swarm optimization, the bat algorithm, as well as new nature inspired paradigms.

In conclusion, the edited book comprises papers on diverse aspects of fuzzy logic augmentation of nature-inspired optimization metaheuristics and their application in areas, such as intelligent control and robotics, pattern recognition, time series prediction, and optimization of complex problems. There are theoretical aspects as well as application papers.

Mexico, May 2014

Oscar Castillo
Patricia Melin

Contents

Part I   Theory

**Fuzzy Logic for Dynamic Parameter Tuning in ACO
and Its Application in Optimal Fuzzy Logic Controller Design**   3
Héctor Neyoy, Oscar Castillo and José Soria

**Fuzzy Classification System Design Using PSO
with Dynamic Parameter Adaptation Through Fuzzy Logic**   29
Frumen Olivas, Fevrier Valdez and Oscar Castillo

**Differential Evolution with Dynamic Adaptation of Parameters
for the Optimization of Fuzzy Controllers**   49
Patricia Ochoa, Oscar Castillo and José Soria

**A New Bat Algorithm with Fuzzy Logic for Dynamical
Parameter Adaptation and Its Applicability to Fuzzy
Control Design**   65
Jonathan Pérez, Fevrier Valdez and Oscar Castillo

**Optimization of Benchmark Mathematical Functions
Using the Firefly Algorithm with Dynamic Parameters**   81
Cinthya Solano-Aragón and Oscar Castillo

**Cuckoo Search via Lévy Flights and a Comparison
with Genetic Algorithms**   91
Maribel Guerrero, Oscar Castillo and Mario García

**A Harmony Search Algorithm Comparison with Genetic
Algorithms**   105
Cinthia Peraza, Fevrier Valdez and Oscar Castillo

Part II Applications

**A Gravitational Search Algorithm for Optimization
of Modular Neural Networks in Pattern Recognition** 127
Beatriz González, Fevrier Valdez, Patricia Melin
and German Prado-Arechiga

**Ensemble Neural Network Optimization Using the Particle Swarm
Algorithm with Type-1 and Type-2 Fuzzy Integration
for Time Series Prediction** 139
Martha Pulido and Patricia Melin

**Clustering Bin Packing Instances for Generating a Minimal Set
of Heuristics by Using Grammatical Evolution** 151
Marco Aurelio Sotelo-Figueroa, Héctor José Puga Soberanes,
Juan Martín Carpio, Héctor J. Fraire Huacuja, Laura Cruz Reyes
and Jorge Alberto Soria Alcaraz

**Comparative Study of Particle Swarm Optimization Variants
in Complex Mathematics Functions** 163
Juan Carlos Vazquez, Fevrier Valdez and Patricia Melin

**Optimization of Modular Network Architectures with a New
Evolutionary Method Using a Fuzzy Combination of Particle Swarm
Optimization and Genetic Algorithms** 179
Fevrier Valdez

Part I
Theory

Fuzzy Logic for Dynamic Parameter Tuning in ACO and Its Application in Optimal Fuzzy Logic Controller Design

Héctor Neyoy, Oscar Castillo and José Soria

Abstract Ant Colony Optimization (ACO) is a population-based constructive metaheuristic that exploits a form of past performance memory inspired by the foraging behavior of real ants. The behavior of the ACO algorithm is highly dependent on the values defined for its parameters. Adaptation and parameter control are recurring themes in the field of bio-inspired algorithms. The present paper explores a new approach of diversity control in ACO. The central idea is to avoid or slow down full convergence through the dynamic variation of a certain parameter. The performance of different variants of the ACO algorithm was observed to choose one as the basis to the proposed approach. A convergence fuzzy logic controller with the objective of maintaining diversity at some level to avoid premature convergence was created. Encouraging results on several travelling salesman problem (TSP) instances and its application to the design of fuzzy controllers, in particular the optimization of membership functions for a unicycle mobile robot trajectory control are presented with the proposed method.

1 Introduction

ACO is inspired by the foraging behavior of ant colonies, and targets discrete optimization problems [1].

The behavior of the ACO algorithm is highly dependent on the values defined for its parameters and has an effect on its convergence. Often these are kept static during the execution of the algorithm. Changing the parameters at runtime, at a given time or depending on the search progress may improve the performance of the algorithm [2–4].

Control the dynamics of convergence to maintain a balance between exploration and exploitation is critical for good performance.

H. Neyoy · O. Castillo (✉) · J. Soria
Tijuana Institute of Technology, Tijuana, Mexico
e-mail: ocastillo@tectijuana.mx

© Springer International Publishing Switzerland 2015
O. Castillo and P. Melin (eds.), *Fuzzy Logic Augmentation of Nature-Inspired Optimization Metaheuristics*, Studies in Computational Intelligence 574,
DOI 10.1007/978-3-319-10960-2_1

3

Early convergence leaves large sections of the search space unexplored. Slow convergence does not concentrate its attention on areas where good solutions were found.

Fuzzy control has emerged as one of the most active and fruitful areas of research in the application of fuzzy set theory.

The methodology of the fuzzy logic controller is useful when processes are too complex for analysis by conventional quantitative techniques or when the available sources of information are interpreted in a qualitatively inaccurate or uncertain way [5].

Determine the correct parameters for the fuzzy logic controller is a complex problem, it is also a task that consumes considerable time. Because of their ability to solve complex NP problems is made use of ACO for the selection of those already mentioned parameters.

There is some interest in using ACO algorithms in mobile robotics [6, 7]. Nowadays robotic automation is an essential part in the manufacturing process. The autonomous navigation of mobile robots is a challenge. A mobile robot can be useful in unattainable goal situations due to geological conditions or where the human being is endangered. So, mobile robotics is an interesting subject for science and engineering.

This paper explores a new method of diversity control in ACO. The central idea is to prevent or stop the total convergence through the dynamic adjustment of certain parameter of the algorithm applied to the design of fuzzy controllers, specifically to the optimization of membership functions of a trajectory controller for a unicycle mobile robot.

The rest of the paper is organized as follows. Section 2 presents an overview of ACO. Section 3 describes a performance analysis on several TSP instances. Section 4 presents a new method of parameter tuning through fuzzy logic, Sect. 5 shows some simulation results in TSP problems, Sect. 6 describes the optimized fuzzy controller, Sect. 7 presents the considerations taken to implement the ACO algorithm in the optimization of membership functions, Sect. 8 describes how the proposal was applied, Sects. 9 and 10 presents simulation results in the membership functions optimization problem, finally Sect. 11 presents some conclusions.

2 Ant Colony Optimization (ACO)

The first ACO algorithm was called Ant System (AS) and its objective was to solve the traveling salesman problem (TSP), whose goal is to find the shortest route to link a number of cities. In each iteration, each ant keeps adding components to build a complete solution, the next component to be added is chosen with respect to a probability that depends on two factors. The pheromone factor that reflects the past experience of the colony and the heuristic factor that evaluates the interest of selecting a component with respect to an objective function. Both factors weighted by the parameters α and β respectively (1).

$$P_{ij}^k = \frac{[\tau_{ij}]^\alpha [\eta_{ij}]^\beta}{\sum_{l \in N_i^k} [\tau_{il}]^\alpha [\eta_{il}]^{\beta_l}}, \quad if\ j \in N_i^k \tag{1}$$

After all ants have built their tours, pheromone trails are updated. This is done by lowering the pheromone value on all arcs by a constant factor (2), which prevents the unlimited accumulation of pheromone trails and allows the algorithm to forget bad decisions previously taken.

$$\tau_{ij} \leftarrow (1 - \rho)\tau_{ij}, \quad \forall (i,j) \in L \tag{2}$$

And by depositing pheromone on the arcs that ants have crossed in its path (3). The better the tour the greater the amount of pheromone that their arcs receive.

$$\tau_{ij} \leftarrow \tau_{ij} + \sum_{k=1}^{n} \Delta\tau_{ij}^k, \quad \forall (i,j) \in L \tag{3}$$

$$\Delta\tau_{ij}^k = \begin{cases} \frac{1}{C^k}, & if\ arc\ (i,j)\ belongs\ to\ T^k; \\ 0, & otherwise; \end{cases}$$

A first improvement on the initial AS, called the elitist strategy for Ant System (EAS). The idea is to provide strong additional reinforcement to the arcs belonging to the best tour found since the start of the algorithm (4) [1].

$$\tau_{ij} \leftarrow \tau_{ij} + \sum_{k=1}^{n} \Delta\tau_{ij}^k + e\Delta\tau_{ij}^{bs}, \quad \forall (i,j) \in L \tag{4}$$

$$\Delta\tau_{ij}^{bs} = \begin{cases} \frac{1}{C^{bs}}, & if\ arc\ (i,j)\ belongs\ to\ T^k \\ 0, & otherwise; \end{cases}$$

Another improvement over AS is the rank-based version of AS (AS_{Rank}). In AS_{rank} each ant deposits an amount of pheromone that decreases with its rank. Additionally, as in EAS, the best-so-far ant always deposits the largest amount of pheromone in each iteration [1].

$$\tau_{ij} \leftarrow \tau_{ij} + \sum_{r=1}^{w-1} (w - r)\Delta\tau_{ij}^k + \Delta\tau_{ij}^{bs} \tag{5}$$

3 Performance Analysis of ACO

To observe the performance of the AS, EAS and AS_{Rank} ACO variants 30 experiments were performed by method for each instance of the examined TSP (Table 1),

Table 1 TSP instances considered

TSP	Number of cities	Best tour length
Burma14	14	3,323
Ulysses22	22	7,013
Berlin52	52	7,542
Eil76	76	538
kroA100	100	21,282

Table 2 Parameters used for each ACO algorithm

ACO	α	β	ρ	m	τ_0
AS	1	2	0.5	n	m/C^{nn}
AS_{Rank}	1	2	0.1	n	$0.5r(r-1)/\rho C^{nn}$
EAS	1	2	0.5	n	$(e+m)/\rho C^{nn}$

$m = n$

$C^{nn} = 20$ for each tsp except burma 14 where $C^{nn} = 10$

EAS $e = 6$

AS_{Rank} $r = w - 1$; $w = 6$

which are in the range of 14 to 100 cities, all extracted from TSPLIB [8], using the parameters recommended by the literature (Table 2) [1].

The behavior of AS and EAS was very similar in all experiments (Tables 3, 4, 5, 6, 7), the performance of the three variants began to worsen by increasing the problem complexity, however AS_{Rank} performance decreased to a lesser extent than their counterparts when the number of cities was greater than 50 (Tables 5, 6, 7).

Since AS_{Rank} had more success finding the minimum and scored lower averages with more complex TSP instances than the other approaches discussed (Figs. 1 and 2). It can be concluded that AS and EAS have better performance when the number of cities is low unlike AS_{Rank} that works best when the number of cities is not so small due to the pheromone deposit mechanism of this approach, where only the w-1 ants with the shorter tours and the ant with the best so far tour are allowed to deposit pheromone. This strategy can lead to a stagnation situation where all the ants follow the same path and construct the same tour [1] as a result of excessive increase in the pheromone trails of suboptimal routes (Figs. 3 and 4).

Table 3 Performance obtained for the TSP instance Burma14

ACO	Best	Average	Successful runs
AS	3,323	3,323	30/30
AS_{Rank}	3,323	3,329	19/30
EAS	3,323	3,323	30/30

Table 4 Performance obtained for the TSP instance Ulysses22

ACO	Best	Average	Successful runs
AS	7,013	7,022	30/30
AS$_{Rank}$	7,013	7,067	19/30
EAS	7,013	7,018	30/30

Table 5 Performance obtained for the TSP instance Berlin52

ACO	Best	Average	Successful runs
AS	7,542	7,557	2/30
AS$_{Rank}$	7,542	7,580	17/30
EAS	7,542	7,554	6/30

Table 6 Performance obtained for the TSP instance Eil76

ACO	Best	Average	Successful runs
AS	547	556	0/30
AS$_{Rank}$	538	543	1/30
EAS	544	555	0/30

Table 7 Performance obtained for the TSP instance KroA100

ACO	Best	Average	Successful runs
AS	22,305	22,483	0/30
AS$_{Rank}$	21,304	21,549	0/30
EAS	22,054	22,500	0/30

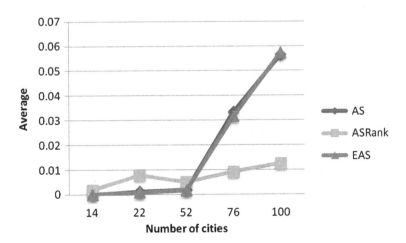

Fig. 1 Average results of each approach discussed

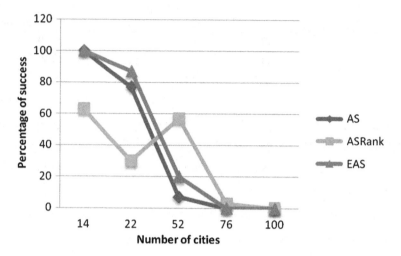

Fig. 2 Percentage of success in finding the global minimum of each approach discussed

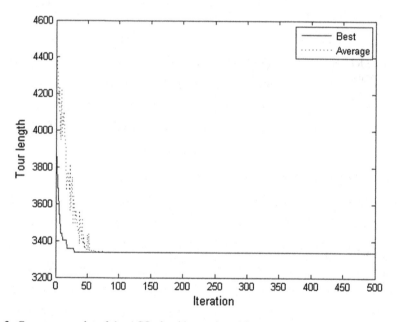

Fig. 3 Convergence plot of the ACO algorithm variant AS_{Rank}

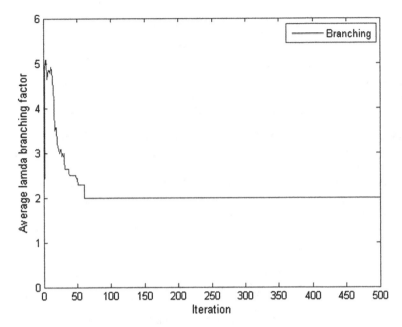

Fig. 4 Behavior of the average lambda branching factor during the execution of the algorithm ACO variant AS_{Rank}

4 Fuzzy Logic Convergence Controller

Due to the obtained results it was decided to use AS_{Rank} as the basis for our proposed ACO variant. The central idea is to prevent or stop the total convergence through the dynamic variation of the alpha parameter.

Alpha has a big effect in the diversity. Is recommended to keep α in the range of $0 < \alpha < 1$ [1].

A value closer to 1 will emphasize better paths but reduce diversity, while lower α will keep more diversity but reduce selective pressure [3].

However, it appears impossible to fix a universally best α. In most approaches it is taken to be 1, so that the selection probability is linear in the pheromone level.

An adaptive parameter control strategy was used; this takes place when there is some form of feedback from the search that is used to determine the direction and/or magnitude of the change to the strategy parameter [9]. In our case, the average lambda branching factor, this factor measures the distribution of the values of the pheromone trails and provides an indication of the size of the search space effectively explored [1].

A convergence fuzzy controller to prevent or delay the full convergence of the algorithm was created (Fig. 5). Fuzzy control can be seen as the translation of external performance specifications and observations of a plant behavior into a rule based linguistic control strategy [5].

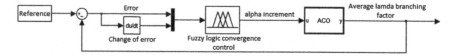

Fig. 5 Block diagram of the proposed system to control the convergence of the ACO algorithm variant AS$_{Rank}$

The objective of the controller is to maintain the average lambda branching factor at a certain level to avoid a premature convergence, so its rules were made to fulfill this purpose (Fig. 6).

The controller uses as inputs the error and change of error (Fig. 7) with respect to an average lambda branching factor reference level and provides as output an increase in the value of parameter alpha (Fig. 8).

If (error is P) and (error_change is P) then (alpha increment is N)
If (error is N) and (error_change is N) then (alpha increment is P)
If (error is P) and (error_change is Z) then (alpha increment is N)
If (error is N) and (error_change is Z) then (alpha increment is P)
If (error is P) and (error_change is N) then (alpha increment is Z)
If (error is N) and (error_change is P) then (alpha increment is Z)
If (error is Z) and (error_change is Z) then (alpha increment is Z)
If (error is Z) and (error_change is N) then (alpha increment is P)
If (error is Z) and (error_change is P) then (alpha increment is N)

Fig. 6 Rules of the proposed fuzzy system to control the convergence of the ACO algorithm

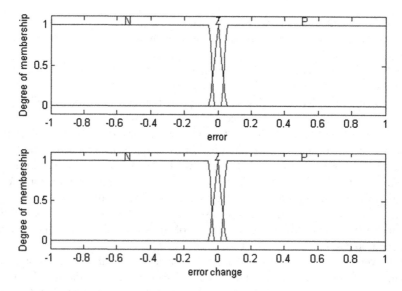

Fig. 7 Membership functions of the input variables of the fuzzy system proposed to control the convergence of the ACO algorithm

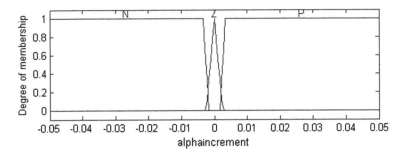

Fig. 8 Membership functions of the output variables of the fuzzy system proposed to control the convergence of the ACO algorithm

5 Simulation in TSP Problems

The controller was able to maintain diversity in a more appropriate level, avoiding the full convergence of the algorithm (Fig. 9).

The same number of experiments that in the above analysis were performed and obtained the following results (Table 8).

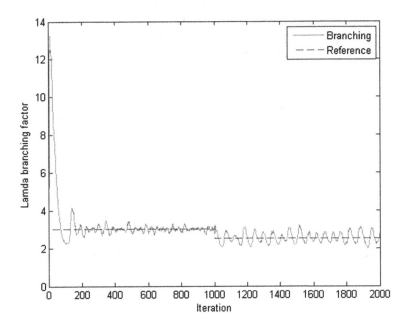

Fig. 9 Behavior of the average lambda branching factor during the execution of the developed approach

Table 8 Performance obtained by the strategy proposed in the instances discussed above

TSP	Best	Average	Successful runs
Burma14	3,323	3,323	30/30
Ulysses22	7,013	7,013	30/30
Berlin52	7,542	7,543	26/30
Eil76	538	539	21/30
KroA100	21,292	21,344	0/30

It was found that the proposed method was able to improve the results of the strategies studied, obtaining lower averages (Fig. 10) and reaching the global minimum on more occasions than the analyzed variants (Fig. 11).

To verify the above in a more formal way a Z test for means of two samples was performed (Table 9).

The 3 ACO variants mentioned above were analyzed in addition to the approach developed in 5 instances of the TSP, 30 experiments were performed for each instance, 150 experiments were made in total of we extracted a 30 data random sample for each method.

With a significance level of 5 % it was found sufficient statistical evidence to claim that the average of AS (Fig. 12a), EAS (Figura 12b) and AS_{Rank} (Fig. 12c) is higher than the obtained for AS_{Rank} + ConvCont in the experiments, this means that our approach improved the performance of the discussed variants on the studied problems, as had been observed in the first analysis.

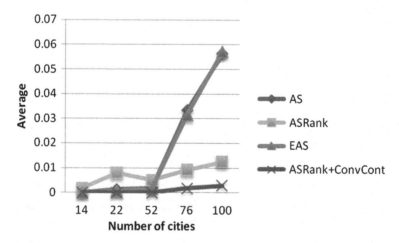

Fig. 10 Average of the results obtained by the proposal and each approach under review

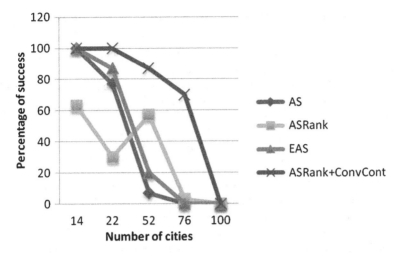

Fig. 11 Percentage of success in finding the global minimum of the proposal and each approach under review

	Case	Null hypothesis (H_0)	Alternative hypothesis (H_a)
Table 9 Null and alternative hypothesis for the statistical hypothesis testing performed for TSP problems	1	$\mu_{AS} \leq \mu_{ASRank+ConvCont}$	$\mu_{AS} > \mu_{ASRank+ConvCont}$
	2	$\mu_{EAS} \leq \mu_{ASRank+ConvCont}$	$\mu_{EAS} > \mu_{ASRank+ConvCont}$
	3	$\mu_{ASRank} \leq \mu_{ASRank+ConvCont}$	$\mu_{ASRank} > \mu_{ASRank+ConvCont}$

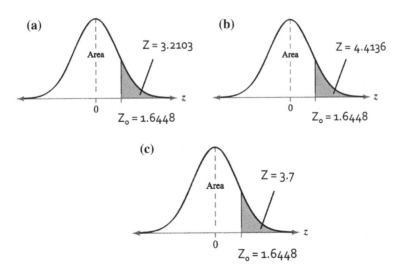

Fig. 12 Results of the statistical hypothesis testing performed for **a** AS vs. AS_{Rank} + ConvCont, **b** EAS vs. AS_{Rank} + ConvCont, **c** AS_{Rank} vs. AS_{Rank} + ConvCont for TSP problems

6 Fuzzy Trajectory Controller for a Unicycle Mobile Robot

It decided to optimize a fuzzy trajectory controller for a unicycle mobile robot to test the developed method in a more complex problem. The control proposal for the mobile robot is: Given a path $q_d(t)$ and a desired orientation, a fuzzy logic controller must be designed to apply an adequate torque τ, such that measured positions $q(t)$ reaches the reference trajectory $q_d(t)$. That is:

$$\lim_{t \to \infty} \| q_d(t) - q(t) \| = 0 \qquad (6)$$

The fuzzy system to optimize [10] is a Takagi-Sugeno type, for simplicity it is decided to modify and convert it into a Mamdani type controller so that the input and output parameters are represented by linguistic variables.

The controller receives as input variables the error in the linear (e_v) and angular (e_w) speed (Fig. 13), that is, the difference between the predefined desired speed and the actual speed of the plant, and as output variables, the right (τ_1) and left (τ_2) torques of said robot (Fig. 14).

The membership functions of the input variables are trapezoidal for the negative (N) and positive (P) linguistic terms, and triangular for the zero (Z) linguistic term. The output variables have three membership functions, negative (N), zero (Z), positive (P) of triangular shape and uses nine fuzzy rules which are shown below (Fig. 15).

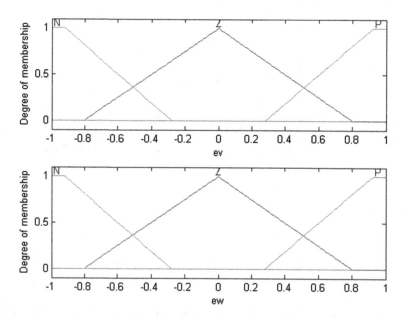

Fig. 13 Membership functions of the fuzzy trajectory controller input variables

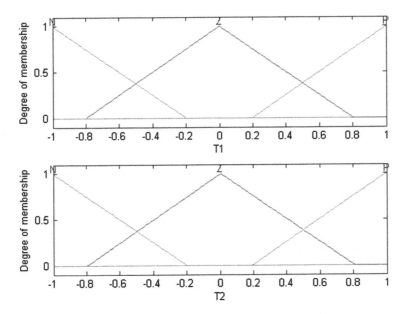

Fig. 14 Membership functions of the fuzzy trajectory controller output variables

If (e_v is N) and (e_w is N) then (τ_1 is N)(τ_2 is N)
If (e_v is N) and (e_w is Z) then (τ_1 is N)(τ_2 is Z)
If (e_v is N) and (e_w is P) then (τ_1 is N)(τ_2 is P)
If (e_v is Z) and (e_w is N) then (τ_1 is Z)(τ_2 is N)
If (e_v is Z) and (e_w is Z) then (τ_1 is Z)(τ_2 is Z)
If (e_v is Z) and (e_w is P) then (τ_1 is Z)(τ_2 is P)
If (e_v is P) and (e_w is N) then (τ_1 is P)(τ_2 is N)
If (e_v is P) and (e_w is Z) then (τ_1 is P)(τ_2 is Z)
If (e_v is P) and (e_w is P) then (τ_1 is P)(τ_2 is P)

Fig. 15 Rules of the of the fuzzy trajectory controller discussed

7 ACO for Membership Functions Optimization

ACO was used to find the membership functions optimal parameters through its adjustment and by the subsequently evaluation of the system.

The parameters a, b, f, j, k corresponding to the membership functions of the input variables remain fixed to simplify the problem. The algorithm will find the optimal values of the parameters c, i in a straightforward manner and, through the optimum position of the intersection points (X1, Y1), (X2, Y2), the value of the parameters d, e, g, h (Fig. 16).

Regarding the membership functions of the output variables, the algorithm will search for the optimum center (b, h, except e that remains fixed for simplicity) and span of each one (a, c, d, f, g, i) (Fig. 17).

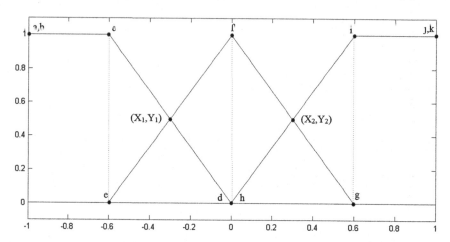

Fig. 16 Membership functions of the input variables of the fuzzy system to control the robot trajectory

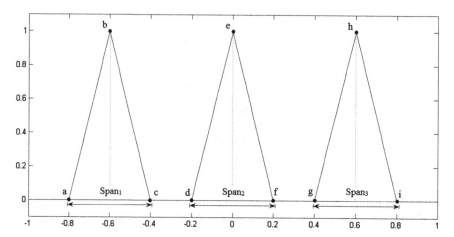

Fig. 17 Membership functions of the output variables of the fuzzy system to control the robot trajectory

The application of ACO to optimize membership functions involves some considerations. First, encode all parameters in a weighted graph. For this purpose we chose a complete graph of 43 nodes to maintain the similarity of the problem with a classical TSP where a minimum Hamiltonian circuit is searched.

The range of each variable was discretized in 22 normalized values in the range [−1 1], and a symmetric data matrix of 43 × 43 with the distance between nodes was created. The parameters of the membership functions of the fuzzy system are obtained through the distance between two nodes using the relations of Tables 10, 11, 12, 13.

Table 10 Relation variable weight for the linear speed error input of the fuzzy system to optimize

Variable	Relation
c	$c = -1 + 0.475 \left(\frac{(d_{1,j_1} + d_{2,j_2})}{2} \right) + 0.475$
X_1	$X_1 = c + \left(m_1 \left(\frac{(d_{3,j_3} + d_{4,j_4})}{2} \right) + c_1 \right)$ $m_1 = -\frac{c}{2}$ $C_1 = -(c + m_1)$
Y_1	$Y_1 = 0.5 \left(\frac{(d_{5,j_5} + d_{6,j_6})}{2} \right) + 0.5$
i	$i = -1 - 0.475 \left(\frac{(d_{11,j_{11}} + d_{12,j_{12}})}{2} \right) + 0.475$
X_2	$X_2 = i + \left(m_2 \left(\frac{(d_{7,j_7} + d_{8,j_8})}{2} \right) + c_2 \right)$ $m_2 = \frac{c}{2}$ $C_2 = i - m_2)$
Y_2	$Y_2 = 0.5 \left(\frac{(d_{9,j_9} + d_{10,j_{10}})}{2} \right) + 0.5$

Table 11 Relation variable weight for the angular speed error input of the fuzzy system to optimize

Variable	Relation
c	$c = -1 + 0.475 \left(\frac{(d_{13,j_{13}} + d_{14,j_{14}})}{2} \right) + 0.475$
X_1	$X_1 = c + \left(m_3 \left(\frac{(d_{15,j_{15}} + d_{16,j_{16}})}{2} \right) + c_3 \right)$ $m_3 = -\frac{c}{2}$ $C_3 = -(c + m_3)$
Y_1	$Y_1 = 0.5 \left(\frac{(d_{17,j_{17}} + d_{18,j_{18}})}{2} \right) + 0.5$
i	$i = 1 - 0.475 \left(\frac{(d_{23,j_{23}} + d_{24,j_{24}})}{2} \right) + 0.475$
X_2	$X_2 = i + \left(m_4 \left(\frac{(d_{19,j_{19}} + d_{20,j_{20}})}{2} \right) + c_4 \right)$ $m_4 = -\frac{c}{2}$ $C_4 = i - m_4)$
Y_2	$Y_2 = 0.5 \left(\frac{(d_{21,j_{21}} + d_{22,j_{22}})}{2} \right) + 0.5$

The next step is to define an appropriate objective function. The objective function represents the quality of the solution, and acts as an interface between the optimization algorithm and the problem considered. The mean square error was used to evaluate the fitness of the fuzzy system.

Table 12 Relation variable weight for the right torque output of the fuzzy system to optimize

Variable	Relation
b	$b = 0.5\left(\frac{d_{25,J_{25}} + d_{26,J_{26}}}{2}\right) - 0.5$
Span$_1$	$span_1 = 0.475\left(\frac{d_{27,J_{27}} + d_{28,J_{28}}}{2}\right) + 0.525$
Span$_2$	$span_2 = 0.475\left(d_{29,J_{29}}\right) + 0.525$
h	$h = 0.5\left(\frac{(d_{30,J_{30}} + d_{31,J_{31}})}{2}\right) + 0.5$
Span$_3$	$span_3 = 0.475\left(\frac{d_{32,J_{32}} + d_{33,J_{33}}}{2}\right) + 0.525$

Table 13 Relation variable weight for the left torque output of the fuzzy system to optimize

Variable	Relation
b	$b = 0.5\left(\frac{d_{34,J_{34}} + d_{35,J_{35}}}{2}\right) - 0.5$
Span$_1$	$span_1 = 0.475\left(\frac{d_{36,J_{36}} + d_{37,J_{37}}}{2}\right) + 0.525$
Span$_2$	$span_2 = 0.475\left(d_{38,J_{38}}\right) + 0.525$
H	$h = 0.5\left(\frac{d_{39,J_{39}} + d_{40,J_{40}}}{2}\right) + 0.5$
Span$_3$	$span_3 = 0.475\left(\frac{d_{41,J_{41}} + d_{42,J_{42}}}{2}\right) + 0.525$

$$MSE = \frac{1}{N}\sum_{K=1}^{N} [y(k) - \tilde{y}(k)]^2 \tag{7}$$

where:

$y(k)$ Reference value at instant k

$\tilde{y}(k)$ Computed output of the system at instant k

N   Number of samples considered

Table 14 Parameters used for each ACO algorithm in the membership function optimization problem

ACO	α	β	ρ	m	τ_0
AS	1	0	0.5	n	m/C^{nn}
AS$_{Rank}$	1	0	0.1	n	$0.5r(r-1)/\rho C^{nn}$
EAS	1	0	0.5	n	$(e+m)/\rho C^{nn}$
AS$_{Rank}$ + CONVCONT	1	0	Dynamic	n	0.1

m = n

C^{nn} = length of a tour generated by a nearest-neighbor heuristic

EAS e = 6

ASRank, AS$_{Rank}$ + CONVCONT: r = w − 1; w = 6

Since the system is responsible for controlling the linear (v) and angular (w) velocities of the plant, the overall error is given by:

$$MSE_v = \frac{1}{N} \sum_{K=1}^{N} [v(k) - \tilde{v}(k)]^2$$

$$MSE_w = \frac{1}{N} \sum_{K=1}^{N} [w(k) - \tilde{w}(k)]^2$$

$$Error_{global} = MSE_v + MSE_w$$

This was used to represent the entire length of each ant generated graph.

8 AS$_{Rank}$ + ConvCont for Membership Functions Optimization

Due to the nature of the problem do not features heuristic information to balance between the influence of the knowledge we have a priori of the problem and the pheromone trails that ants have generated, thus the dynamic variation of the parameter alpha had a null effect on the convergence of the algorithm when applied to the optimization of membership functions (Fig. 18).

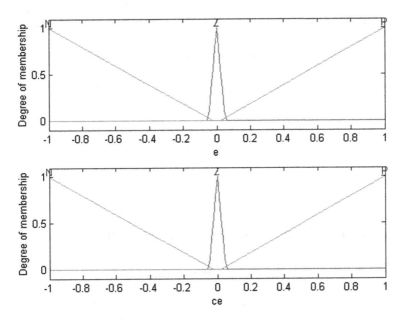

Fig. 18 Membership functions of the input variables of the fuzzy system proposed to control the convergence of the ACO algorithm without heuristic information

If (error is P) and (error_change is P) then (Δ_u^{bs} N) (Δ_u P) (ρ^{bs} P) (ρN)

If (error is N) and (error_change is N) then (Δ_u^{bs} P) (Δ_u N) (ρ^{bs} N) (ρP)

If (error is P) and (error_change is Z) then (Δ_u^{bs} N) (Δ_u P) (ρ^{bs} P) (ρN)

If (error is N) and (error_change is Z) then (Δ_u^{bs} N) (Δ_u P) (ρ^{bs} P) (ρN)

If (error is P) and (error_change is N) then (Δ_u^{bs} N) (Δ_u P) (ρ^{bs} P) (ρN)

If (error is N) and (error_change is P) then (Δ_u^{bs} N) (Δ_u P) (ρ^{bs} P) (ρN)

If (error is Z) and (error_change is Z) then (Δ_u^{bs} N) (Δ_u P) (ρ^{bs} P) (ρN)

If (error is Z) and (error_change is N) then (Δ_u^{bs} N) (Δ_u P) (ρ^{bs} P) (ρN)

If (error is Z) and (error_change is P) then (Δ_u^{bs} N) (Δ_u P) (ρ^{bs} P) (ρN)

Fig. 19 Rules of the proposed fuzzy system to control the convergence of the ACO algorithm without heuristic information

It was decided to continue with the same strategy of convergence control, but this time by varying the evaporation rate (ρ) and the weight to be given to the amount of pheromone that each ant leaves on its trail (w) to control diversity, so a fuzzy system was implemented for this task.

The controller uses as inputs the error (e) and change of error (ce) with respect to an average lambda branching factor reference level (Fig. 18) and provides as output the evaporation rate corresponding to arcs which belong (ρ^{bs}) and do not belong (ρ) to the best so far tour in addition to an increase in the weight that is given to the pheromone increment of the arcs that form part of the best so far tour (u^{bs}) and the remaining arcs (u) in AS$_{Rank}$ (Fig. 19).

Again the rules were created with the intention to keep the average lambda branching factor at some level to slow the convergence process and are shown (Fig. 20):

Thus Eqs. 2 and 4 corresponding to the evaporation and pheromone deposit process in AS$_{Rank}$ become:

$$\tau_{ij}^{bs} \leftarrow (1 - \rho^{bs})\tau_{ij}^{bs}, \quad \forall(i,j) \in T^{bs}$$

$$\tau_{ij}^{bs} \leftarrow (1 - \rho)\tau_{ij}, \quad \forall(i,j) \notin T^{bs}$$

$$\tau_{ij} \leftarrow \tau_{ij} + \sum_{r=1}^{w-1} \frac{(w - r)(u)}{(w - 1)}\Delta\tau_{ij}^r + u^{bs}\Delta_{ij}^{bs}$$

$$\Delta\tau_{ij}^r = \frac{1}{C^r} \ y \ \Delta\tau_{ij}^{bs} = \frac{1}{C^{bs}}$$

9 Simulation in Membership Functions Optimization Problem

The model of the mobile robot and the path used in the simulations performed by the ACO algorithm are defined in [10].

The approach described in previous section was able to maintain diversity in the required level (Fig. 21) unlike the convergence controller that was tested in the Sect. 5.

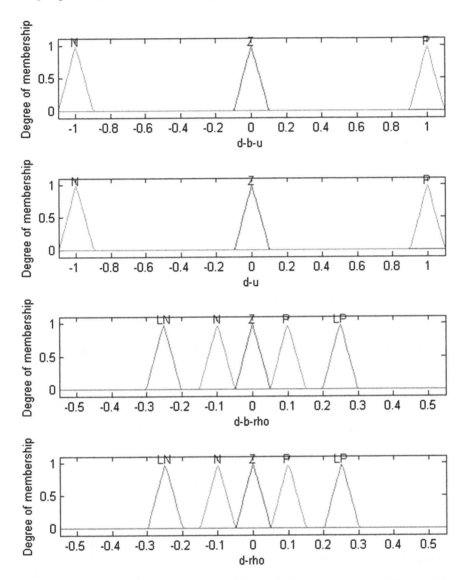

Fig. 20 Membership functions of the output variables of the fuzzy system proposed to control the convergence of the ACO algorithm without heuristic information

30 experiments were performed by approach (Table 15) to compare the performance of classical approaches with the developed proposal. Using the following parameters (Table 14).

With the exception of AS_{Rank}, the average simulation results obtained were very similar. The proposal got the lowest average but despite that was EAS which generated the lowest MSE controller (Fig. 23) and therefore more accurate trajectory (Fig. 22).

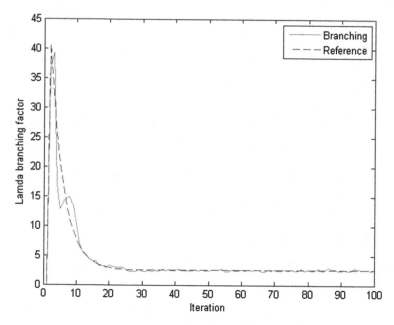

Fig. 21 Behavior of the average lambda branching factor during the execution of the developed approach to control the convergence of the ACO algorithm without heuristic information

Table 15 Results obtained by the proposal and each approach under review algorithm in the memerbership function optimization problem

ACO	Best	Average
AS	0.0015	0.0172
EAS	0.00013	0.0161
AS_{Rank}	0.00015	0.0572
AS_{Rank} + CONVCONT	0.00029	0.0131

It is difficult to determine whether the proposal exceeded the classical approaches with the above analysis, so a Z test for two samples means was performed to come to a conclusion (Table 16).

No statistical evidence was found with a significance level of 5 % that the average of AS or EAS is greater than the average of AS_{Rank} + CONVCONT (Figs. 24a, b).

With a significance level of 5 %, only statistical evidence that the average of the results of simulations of AS_{Rank} is greater than AS_{Rank} + CONVCONT was found (Fig. 24c), that is, the proposal was only able to outperform the AS_{Rank} variant.

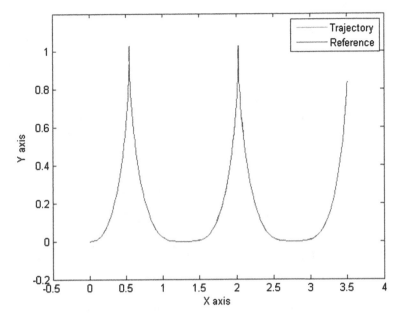

Fig. 22 Trajectory obtained by the best generated controller

10 AS$_{Rank}$ + ConvCont vs. S-ACO

The results obtained with the developed proposal were compared with the obtained by [6] who attacked the same membership function optimization problem for the same fuzzy trajectory controller and unicycle mobile robot model, the difference lies in S-ACO as strategy used to solve the problem and the directed graph of 12 nodes chosen to represent it.

At first glance it can be observed that the best result AS$_{Rank}$ + CONVCONT was significantly lower than S-ACO as well as the average of the results obtained in the experiments (Table 17), this is reflected in the path generated by each controller (Fig. 25), therefore we conclude that its performance is higher.

To support the above a t-test for means of two samples was performed, for which it took a random sample of 10 experiments per technique to compare their performance.

The null hypothesis claims that the average of S-ACO is less than or equal to AS$_{Rank}$ + CONVCONT.

Since t is located at the rejection zone with a significance level of 5 % and 9 degrees of freedom there is sufficient statistical evidence to prove that the average of S-ACO is greater than AS$_{Rank}$ + ConvCont (Fig. 26), that is, the developed approach outperformed the method used by [6] and therefore likewise AS and EAS by the Sect. 9 analysis.

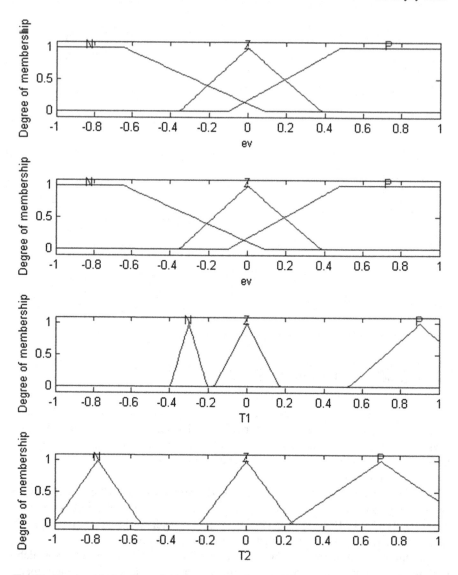

Fig. 23 Membership functions of the best generated controller

Table 16 Null and alternative hypothesis for the statistical hypothesis testing performed for membership function optimization problem

Case	Null hypothesis (H_0)	Alternative hypothesis (H_a)
1	$\mu_{AS} \leq \mu_{ASRank+ConvCont}$	$\mu_{AS} > \mu_{ASRank+ConvCont}$
2	$\mu_{EAS} \leq \mu_{ASRank+ConvCont}$	$\mu_{EAS} > \mu_{ASRank+ConvCont}$
3	$\mu_{ASRank} \leq \mu_{ASRank+ConvCont}$	$\mu_{ASRank} > \mu_{ASRank+ConvCont}$

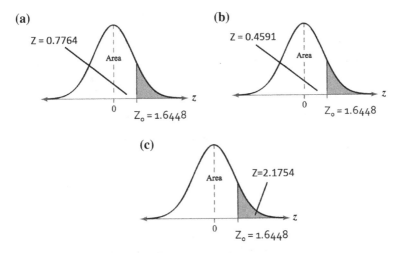

Fig. 24 Results of the statistical hypothesis testing performed for **a** AS vs. AS_{Rank} + ConvCont, **b** EAS vs. AS_{Rank} + ConvCont, **c** AS_{Rank} vs. AS_{Rank} + ConvCont for membership functions optimization problem

Table 17 Performance obtained by AS_{Rank} + CONV-CONT and S-ACO in the membership function optimization problem

ACO	Best	Average
AS_{Rank} + CONVCONT	0.00029	0.0131
S-ACO	0.0982	0.1199

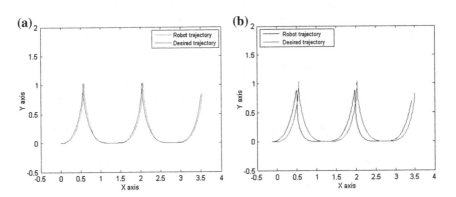

Fig. 25 Trajectories generated by the controller obtained by the best of experiments performed with: **a** AS_{Rank} + ConvCont, **b** S-ACO

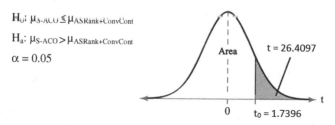

Fig. 26 Results of the statistical hypothesis testing performed for a S-ACO vs. AS$_{Rank}$ + ConvCont

11 Conclusions

Maintaining diversity is important for good performance in the ACO algorithm. An adaptive control strategy of the parameter alpha for this purpose was used, which was embodied in a diversity fuzzy controller which allows avoiding or delaying the total convergence and thereby controlling the exploration and exploitation capabilities of the algorithm.

The strategy was compared with 3 variants of the ACO algorithm on several instances of the TSP taken from TSPLIB. An improvement was observed by dynamically changing the parameter alpha value, as was seen in the statistical analysis performed, where our approach outperforms the classical strategies.

It was found that the parameter alpha is not the most appropriate when there is no heuristic information to guide the search as was the case with the optimization of membership functions, since it is not possible to balance between the previous knowledge of the problem and by the generated by the algorithm itself during its execution and thus control the convergence of the algorithm. So it was decided to continue with the same strategy for this kind of problem, but varying the evaporation rate and the weight that is given to the amount of pheromone which each ant deposited, what allowed controlling the convergence of the algorithm without heuristic information. This modification improved the performance of AS$_{Rank}$, however since this variant scored the lowest performance, is probably not the most appropriate in these cases.

The formulated strategy was outperformed by AS and EAS in the membership functions optimization problem but managed to outperform the method developed in [6], so it was concluded that the improvement could not come from the convergence control made and is attributed to the way in which the problem was encoded.

As future work it intends to apply convergence control to other variants of ACO algorithm. Modify the reference, and thus diversity in an intelligent way, depending of the search progress or some other performance measure. Look for heuristic information relevant to the membership functions optimization problem that drives the search process in early iterations of the algorithm, making it possible to use the strategy of dynamic variation of the parameter alpha and an analysis in presence of noise of the generated controller by ACO algorithm.

References

1. Dorigo, M., Stutzle, T.: Ant Colony Optimization. MIT Press, Cambridge (2004)
2. Merkle, D., Middendorf, M.: Prospects for dynamic algorithm control: lessons from the phase structure of ant scheduling algorithms. In: Heckendorn R.B. (ed) Proceedings of the 2000 Genetic and Evolutionary Computation Conference—Workshop Program. Workshop "The Next Ten Years of Scheduling Research", Morgan Kaufmann Publishers (2001)
3. Merkle, D., Middendorf, M., Schmeck, H.: Ant colony optimization for resource-constrained project scheduling. IEEE Trans. Evol. Comput. **6**, 333–346 (2002)
4. B Meyer, Convergence control in ACO. Genetic and Evolutionary Computation Conference (GECCO), Seattle (2004)
5. Yen, J., Langari, R.: Fuzzy Logic: Intelligence, Control and Information. Prentice Hall, Englewood Cliffs (1999)
6. Castillo, O., Martinez-Marroquin, R., Melin, P., Valdez, F., Soria, J.: Comparative study of bio inspired algorithms applied to the optimization of type-1 and type-2 fuzzy controllers for an autonomous mobile robot. Inf. Sci. (2010, in press)
7. M Mohamad, W Dunningan, Ant colony robot motion planning. In: International Conference on Computer as a Tool, EUROCON, Serbia and Montenegro (2005)
8. Reinelt, G.: *TSPLIB*. University of Heidelberg, http://comopt.ifi.uni-heidelberg.de/software/TSPLIB95/
9. Eiben, A.E., Hinterding, R., Michalewicz, Z.: Parameter control in evolutionary algorithms. IEEE Trans. Evol. Comput. **3**(2), 124–141 (1999)
10. Martínez, R., Castillo, O., Aguilar, L.: Optimization of interval type-2 fuzzy logic controllers for a perturbed autonomous wheeled mobil robot using genetic algorithms. Inf. Sci. **179**, 2158–2174 (2009)
11. Aceves, A., Aguilar, J.: A simplified version of Mamdani's fuzzy controller: the natural logic controller. IEEE Trans. Fuzzy Syst. **14**(1), 16–30 (2006)
12. Alcalá-Fdez, J., Alcalá, R., Gacto, M.J., Herrera, F.: Learning the membership function contexts forming fuzzy association rules by using genetic algorithms. Fuzzy Sets Syst. **160**(7), 905–921 (2009)
13. Bloch, A.M., Drakunov, S.: Tracking in non-Holonomic dynamic system via sliding modes. In: Proceedings of IEEE Conference on Decision and Control, Brighton, pp. 1127–1132 (1991)
14. Campion, G., Bastin, G., D'Andrea-Novel, B.: Structural properties and classification of kinematic and dynamic models of wheeled mobile robots. IEEE Trans. Robot. Autom. **12**(1), 47–62 (1996)
15. Chusanapiputt, S., Nualhong, D., Jantarang, S., Phoomvuthisarn, S.: Selective self-adaptive approach to ant system for solving unit commitment problem. In: Cattolico M., et al. (eds.) GECCO 2006, pp. 1729–1736. ACM Press, New York (2006)
16. Chwa, D.: Sliding-mode tracking control of nonholonomic wheeled mobile robots in polar coordinates. IEEE Trans. Control Syst. Tech. **12**(4), 633–644 (2004)
17. Fukao, T., Nakagawa, H., Adachi, N.: Adaptive tracking control of a nonholonomic mobile robot. IEEE Trans. Rob. Autom. **16**(5), 609–615 (2000)
18. Hao, Z., Huang, H., Qin, Y., Cai, R.: An ACO algorithm with adaptive volatility rate of pheromone trail. Lect. Notes Comput. Sci. **4490**, 1167–1170 (2007)
19. Hong, T., Chen, C., Wu, Y., Lee, Y.: Using divide-and-conquer GA strategy in fuzzy data mining. IEEE International Symposium on Fuzzy Systems, pp. 116–121. Budapest, Hungary (2004)
20. Hong, T.P., Tung, Y.F., Wang, S.L., Wu, M.T., Wu, Y.L.: Extracting membership functions in fuzzy data mining by ant colony systems. Extracting Membership Functions Fuzzy Data Min. Ant Colony Syst. **7**, 3979–3984 (2008)
21. Ishikawa, S., A method of indoor mobile robot navigation by fuzzy control. In: Proceedings International Conference Intelligence Robotics System, pp. 1013–1018. Osaka, Japan (1991)

22. Kanayama, Y., Kimura, F., Miyazaki, T., Noguchi, A.: Stable tracking control method for a non-holonomic mobile robot. Proceedings IEEE/RSJ International Workshop on Intelligent Robots and Systems, pp. 1236–1241. Osaka, Japan (1991)
23. Khalil, H.: Nonlinear Systems, 3rd edn. Prentice Hall, New York (2002)
24. Kulkarni, A.: Computer Vision and Fuzzy-Neural Systems. Prentice Hall, Englewood Cliffs (2001)
25. Lee, T.H., Leung, F.H.F., Tam, P.K.S.: Position Control for Wheeled Mobile Robot Using a Fuzzy Controller, pp 525–528. IEEE (1999)
26. Lee, T.C., Tai, K.: Tracking control of unicycle-modeled mobile robots using a saturation feedback controller. IEEE Trans. Control Syst. Technol. 9(2), 305–318 (2001)
27. Li, Y., Li, W.: Adaptive ant colony optimization algorithm based on information entropy, Foundation and application. Fundamenta Informaticae (2007)
28. Liberzon, D.: Switching in Systems and Control, Bikhauser (2003)
29. Man, K.F., Tang, K.S., Kwong, S.: Genetic Algorithms: Concepts and Designs. Springer, Berlin (2000)
30. Mendel, J.: Uncertain Rule-Based Fuzzy Logia Systems, Introduction and New Directions. Prentice-Hall, Englewood Cliffs (2001)
31. W Nelson, I Cox, Local Path Control for an Autonomous Vehicle. Proceedings of IEEE Conference on Robotics and Automation, pp. 1504–1510 (1988)
32. Paden, B., Panja, R.: Globally asymptotically stable PD + controller for robot manipulator. Int. J. Control 47(6), 1697–1712 (1988)
33. Parvinder, K., Shakti, K., Amarpartap, S.: Optimization of membership functions based on ant colony algorithm. Int. J. Comput. Sci. Inf. Secur. 10(4), 38–45 (2012)
34. Pawlowski, S., Dutkiewicz, P., Kozlowski, K., Wroblewski, W.: Fuzzy logic implementation in mobile robot control. In: 2nd Workshop on Robot Motion and Control, pp 65–70 (2001)
35. Stützle, T., López-Ibañez, M., Pellegrini, P., Maur, M., Montes de Oca, M.A., Birattar, M., Dorigo, M.: Parameter adaptation in ant colony optimization. In: Hamadi, Y., Monfroy, E., Saubion, F. (eds.) Autonomous Search. Springer, Berlin (2012)
36. Takagi, T., Sugeno, M.: Fuzzy identification of systems and its application to modeling and control. IEEE Trans. Syst. Man Cybern. 15(1), 116–132 (1985)
37. Tsai, C.C., Lin, H.H., Lin, C.C.: Trajectory tracking control of a laser-guided wheeled mobile robot. In: Proceedings IEEE International Conferences on Control Applications, pp. 1055–1059. Taipei (2004)
38. Ulyanov, S.V., Watanabe, S., Ulyanov, V.S., Yamafuji, K., Litvintseva, L.V., Rizzotto, G.G.: Soft computing for the intelligent robust control of a robotic unicycle with a new physical measure for mechanical controllability. Soft Comput. 2, 73–88 (1998). (Springer)
39. Valdez, F., Melin, P., Castillo, O.: An improved evolutionary method with fuzzy logic for combining particle swarm optimization and genetic algorithms. Appl. Soft Comput. 11, 2625–2632 (2011)
40. Ye, W., Ma, D., Fan, H.: Path planning for space robot based on the self-adaptive ant colony algorithm. Proceedings of the 1st International Symposium on Systems and Control in Aerospace and Astronautics ISSCAA, IEEEXplore (2006)

Fuzzy Classification System Design Using PSO with Dynamic Parameter Adaptation Through Fuzzy Logic

Frumen Olivas, Fevrier Valdez and Oscar Castillo

Abstract In this paper a new method for dynamic parameter adaptation in particle swarm optimization (PSO) is proposed. PSO is a metaheuristic inspired in social behaviors, which is very useful in optimization problems. In this paper we propose an improvement to the convergence and diversity of the swarm in PSO using fuzzy logic. Simulation results show that the proposed approach improves the performance of PSO.

Keywords Fuzzy logic · Particle swarm optimization · Dynamic parameter adaptation · Fuzzy classifier · Fuzzy classification system

1 Introduction

Fuzzy logic or multi-valued logic is based on fuzzy set theory proposed by Zadeh [14], which helps us in modeling knowledge, through the use of if-then fuzzy rules.

The fuzzy set theory provides a systematic calculus to deal with linguistic information, and that improves the numerical computation by using linguistic labels stipulated by membership functions [7].

Particle swarm optimization (PSO) that was introduced by Kennedy and Eberhart in 1995 [9, 10], maintains a swarm of particles and each particle represents a possible solution. These particles "fly" through a multidimensional search space, where the position of each particle is adjusted according to your own experience and that of its neighbors [4].

PSO has recently received many improvements and applications. Most of the modifications to PSO are to improve convergence and to increase the diversity of the swarm [4]. So in this paper we propose an improvement to the convergence and diversity of PSO through the use of fuzzy logic. Basically, fuzzy rules are used to

F. Olivas · F. Valdez · O. Castillo (✉)
Tijuana Institute of Technology, Tijuana, Mexico
e-mail: ocastillo@tectijuana.mx

© Springer International Publishing Switzerland 2015   29
O. Castillo and P. Melin (eds.), *Fuzzy Logic Augmentation of Nature-Inspired Optimization Metaheuristics*, Studies in Computational Intelligence 574,
DOI 10.1007/978-3-319-10960-2_2

control the key parameters in PSO to achieve the best possible dynamic adaptation of these parameter values.

The rest of the paper is organized as follows. Section 2 describes the proposed methodology. Section 3 shows how the experiments were performed with the proposed method and the simple method using the benchmark functions defined in Sect. 2. Section 4 shows how to perform the statistical comparison with all its parameters and analysis of results. Section 5 shows the design of fuzzy classifier. Section 6 shows the methodology to follow for the design of fuzzy classifier. Section 7 shows how the experiments were performed with the proposed method and the simple method in the design of fuzzy classifier. Section 8 shows how to perform the statistical comparison with all its parameters and analysis of results. Section 9 shows the conclusions of the design of fuzzy classifier design. Finally, the conclusions of this paper are presented.

2 Methodology for Parameter Adaptation

The dynamics of PSO is defined by Eqs. 1 and 2, which are the equations to update the position and velocity of the particle, respectively.

$$x_i(t+1) = x_i(t) + v_i(t+1) \tag{1}$$

$$v_{ij}(t+1) = v_{ij}(t) + c_1 r_1(t)\lfloor y_{ij}(t) - x_{ij}(t)\rfloor + c_2 r_{2j}(t)\lfloor \hat{y}_j(t) - x_{ij}(t)\rfloor \tag{2}$$

Parameters c1 and c2 were selected to be adjusted using fuzzy logic, since those parameters account for the movement of the particles.

The parameter c1 or cognitive factor represents the level of importance given the particle to its previous positions.

The parameter c2 or social factor represents the level of importance that the particle gives the best overall position.

Based on the literature [4] the recommended values for c1 and c2 must be in the range of 0.5 and 2.5, plus it is also suggested that changing the parameters c1 and c2 dynamically during the execution of this algorithm can produce better results.

In addition it is also found that the algorithm performance measures, such as: diversity of the swarm, the average error at one point in the execution of the algorithm, the iterations themselves, needs to be considered to run the algorithm, among others. In our work all the above are taken in consideration for the fuzzy systems to modify the parameters c1 and c2 dynamically changing these parameters in each iteration of the algorithm.

For measuring the iterations of the algorithm, it was decided to use a percentage of iterations, i.e. when starting the algorithm the iterations will be considered "low", and when the iterations are completed it will be considered "high" or close to 100 %. To represent this idea we use Eq. 3.

$$Iteration = \frac{Current\ Iteration}{Maximum\ of\ Iterations} \tag{3}$$

The diversity measure is defined by Eq. 4, which measures the degree of dispersion of the particles, i.e. when the particles are closer together there is less diversity as well as when particles are separated then diversity is high. As the reader will realize the equation of diversity can be considered as the average of the Euclidean distances between each particle and the best particle.

$$Diversity(S(t)) = \frac{1}{n_s} \sum_{i=1}^{n_s} \sqrt{\sum_{j=1}^{n_x} \left(x_{ij}(t) - \bar{x}_j(t)\right)^2} \tag{4}$$

The error measure is defined by Eq. 5, which measures the difference between the swarm and the best particle, by averaging the difference between the fitness of each particle and the fitness of the best particle.

$$Error = \frac{1}{n_s} \sum_{i=1}^{n_s} \left(Fitness(x_i) - MinF\right) \tag{5}$$

Therefore for designing the fuzzy systems, which dynamically adjust the parameters of c1 and c2, the three measures described above were considered as inputs. It is obvious that for each fuzzy system the outputs are c1 and c2.

In regards to the inputs of the fuzzy systems, the iteration variable has by itself a defined range of possible values which range from 0 to 1 (0 is 0 % and 1 is the 100 %), but with the diversity and the error, we perform a normalization of the values of these to have values between 0 and 1. Equation 6 shows how the normalization of diversity is performed and Eq. 7 shows how the normalization of the error is obtained.

$$Diver\,Norm = \begin{cases} if\ Min\,Diver = Max\,Diver\{Diver\,Norm = 0 \\ if\ Min\,Diver \neq Max\,Diver\{Diver\,Norm = Fn\,Norm \end{cases}$$
$$Fn\,Norm = \frac{Diversity - Min\,Diver}{Max\,Diver - Min\,Diver} \tag{6}$$

Equation 6 shows two conditions for the normalization of diversity, the first provides that where the maximum Euclidean distance is equal to the minimum Euclidean distance, this means that the particles are exactly in the same position so there is no diversity. The second condition deals with the cases with different Euclidean distances.

$$Error\,Norm = \begin{cases} if\ MinF = MaxF\{Error\,Norm = 1 \\ if\ MinF \neq MaxF\{Error\,Norm = \frac{Error - MinF}{MaxF - MinF} \end{cases} \tag{7}$$

Equation 7 shows two conditions to normalize the error, the first one tells us that when the minimum fitness is equal to the maximum fitness, then the error will be 1; this is because the particles are close together. The second condition deals with the cases with different fitness.

The design of the input variables can be appreciated in Figs. 1, 2 and 3, which show the inputs iteration, diversity, and error respectively, each input is granulated into three triangular membership functions.

For the output variables, as mentioned above, the recommended values for c1 and c2 are between 0.5 and 2.5, so that the output variables were designed using this range of values. Each output is granulated in five triangular membership functions, the design of the output variables can be seen in Figs. 4 and 5, c1 and c2 respectively.

Fig. 1 Input 1: iteration

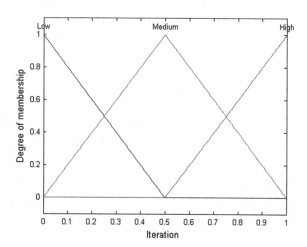

Fig. 2 Input 2: diversity

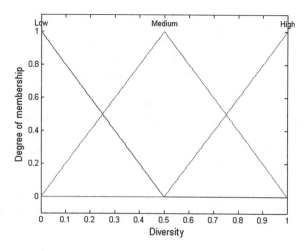

Fig. 3 Input 3: error

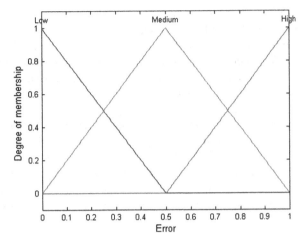

Fig. 4 Output 1: c1

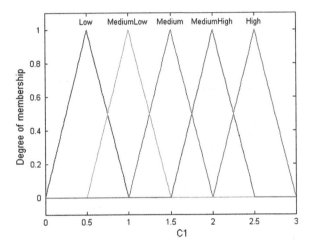

Fig. 5 Output 2: c2

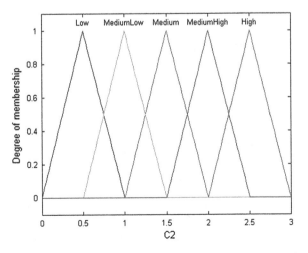

Having defined the possible input variables, it was decided to combine them to generate different fuzzy systems for dynamic adjustment of c1 and c2. Based on the combinations of possible inputs, there were seven possible fuzzy systems, but it was decided to consider only the systems that have more inputs (since we previously considered fuzzy systems with only a single input), so that eventually there were three fuzzy systems which are defined below.

The first fuzzy system has iteration and diversity as inputs, which is shown in Fig. 6. The second fuzzy system has iteration and error as inputs and is shown in Fig. 7. The third fuzzy system has iteration, diversity, and error as inputs, as shown in Fig. 8.

To design the rules of each fuzzy system, it was decided that in early iterations the PSO algorithm must explore and eventually exploit. Taking into account other variables such as diversity, for example, when diversity is low, that is, that the

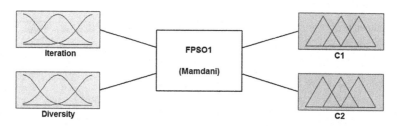

Fig. 6 First fuzzy system

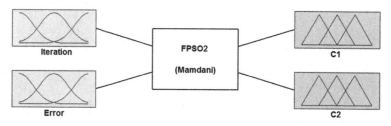

Fig. 7 Second fuzzy system

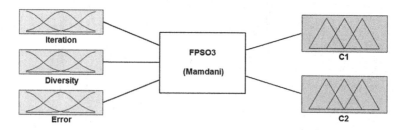

Fig. 8 Third fuzzy system

1. If (Iteration is Low) and (Diversity is Low) then (C1 is High)(C2 is Low)
2. If (Iteration is Low) and (Diversity is Medium) then (C1 is MediumHigh)(C2 is Medium)
3. If (Iteration is Low) and (Diversity is High) then (C1 is MediumHigh)(C2 is MediumLow)
4. If (Iteration is Medium) and (Diversity is Low) then (C1 is MediumHigh)(C2 is MediumLow)
5. If (Iteration is Medium) and (Diversity is Medium) then (C1 is Medium)(C2 is Medium)
6. If (Iteration is Medium) and (Diversity is High) then (C1 is MediumLow)(C2 is MediumHigh)
7. If (Iteration is High) and (Diversity is Low) then (C1 is Medium)(C2 is High)
8. If (Iteration is High) and (Diversity is Medium) then (C1 is MediumLow)(C2 is MediumHigh)
9. If (Iteration is High) and (Diversity is High) then (C1 is Low)(C2 is High)

Fig. 9 Rules for fuzzy system 1

1. If (Error is Low) and (Iteration is Low) then (C1 is Low)(C2 is Medium)
2. If (Error is Low) and (Iteration is Medium) then (C1 is MediumLow)(C2 is MediumHigh)
3. If (Error is Low) and (Iteration is High) then (C1 is Low)(C2 is High)
4. If (Error is Medium) and (Iteration is Low) then (C1 is MediumLow)(C2 is MediumHigh)
5. If (Error is Medium) and (Iteration is Medium) then (C1 is Medium)(C2 is Medium)
6. If (Error is Medium) and (Iteration is High) then (C1 is Medium)(C2 is High)
7. If (Error is High) and (Iteration is Low) then (C1 is High)(C2 is MediumLow)
8. If (Error is High) and (Iteration is Medium) then (C1 is MediumHigh)(C2 is Medium)
9. If (Error is High) and (Iteration is High) then (C1 is High)(C2 is Low)

Fig. 10 Rules for fuzzy system 2

particles are close together, we must use exploration, and when diversity is high we must use exploitation.

The rules for each fuzzy system are shown in Figs. 9, 10 and 11, for the fuzzy systems 1, 2 and 3, respectively.

Also for the comparison of the proposed method with respect to the PSO without parameter adaptation, we considered benchmark mathematical functions, defined in [6, 11], which are 27 in total, and in each we must find the parameters that give us the global minimum of each function. In Fig. 12 there is a sample of the functions that are used.

As indicated in Fig. 12 we only considered functions of one or two dimensions for the experiments.

So that once defined the fuzzy systems that dynamically adjust the parameters of PSO, and defined the problem in which it applies (Benchmark mathematical functions), the proposal is as shown in Fig. 13, where we can notice that c1 and c2 parameters are adjusted by a fuzzy system, and in turn this "fuzzy PSO" searches for the optimal parameters for the Benchmark mathematical functions.

1. If (Iteration is Low) and (Diversity is Low) and (Error is Low) then (C1 is MediumLow)(C2 is Low)
2. If (Iteration is Low) and (Diversity is Low) and (Error is Medium) then (C1 is MediumHigh)(C2 is Low)
3. If (Iteration is Low) and (Diversity is Low) and (Error is High) then (C1 is High)(C2 is Low)
4. If (Iteration is Low) and (Diversity is Medium) and (Error is Low) then (C1 is Medium)(C2 is Medium)
5. If (Iteration is Low) and (Diversity is Medium) and (Error is Medium) then (C1 is MediumHigh)(C2 is MediumLow)
6. If (Iteration is Low) and (Diversity is Medium) and (Error is High) then (C1 is MediumHigh)(C2 is Low)
7. If (Iteration is Low) and (Diversity is High) and (Error is Low) then (C1 is Low)(C2 is MediumLow)
8. If (Iteration is Low) and (Diversity is High) and (Error is Medium) then (C1 is Medium)(C2 is Medium)
9. If (Iteration is Low) and (Diversity is High) and (Error is High) then (C1 is MediumLow)(C2 is Low)
10. If (Iteration is Medium) and (Diversity is Low) and (Error is Low) then (C1 is Medium)(C2 is Medium)
11. If (Iteration is Medium) and (Diversity is Low) and (Error is Medium) then (C1 is MediumHigh)(C2 is MediumLow)
12. If (Iteration is Medium) and (Diversity is Low) and (Error is High) then (C1 is MediumHigh)(C2 is Low)
13. If (Iteration is Medium) and (Diversity is Medium) and (Error is Low) then (C1 is MediumLow)(C2 is MediumHigh)
14. If (Iteration is Medium) and (Diversity is Medium) and (Error is Medium) then (C1 is Medium)(C2 is Medium)
15. If (Iteration is Medium) and (Diversity is Medium) and (Error is High) then (C1 is MediumHigh)(C2 is MediumLow)
16. If (Iteration is Medium) and (Diversity is High) and (Error is Low) then (C1 is Low)(C2 is MediumHigh)
17. If (Iteration is Medium) and (Diversity is High) and (Error is Medium) then (C1 is MediumLow)(C2 is MediumHigh)
18. If (Iteration is Medium) and (Diversity is High) and (Error is High) then (C1 is Medium)(C2 is Medium)
19. If (Iteration is High) and (Diversity is Low) and (Error is Low) then (C1 is Low)(C2 is MediumLow)
20. If (Iteration is High) and (Diversity is Low) and (Error is Medium) then (C1 is Medium)(C2 is Medium)
21. If (Iteration is High) and (Diversity is Low) and (Error is High) then (C1 is MediumLow)(C2 is Low)
22. If (Iteration is High) and (Diversity is Medium) and (Error is Low) then (C1 is Low)(C2 is MediumHigh)
23. If (Iteration is High) and (Diversity is Medium) and (Error is Medium) then (C1 is MediumLow)(C2 is MediumHigh)
24. If (Iteration is High) and (Diversity is Medium) and (Error is High) then (C1 is Medium)(C2 is Medium)
25. If (Iteration is High) and (Diversity is High) and (Error is Low) then (C1 is Low)(C2 is High)
26. If (Iteration is High) and (Diversity is High) and (Error is Medium) then (C1 is Low)(C2 is MediumHigh)
27. If (Iteration is High) and (Diversity is High) and (Error is High) then (C1 is Low)(C2 is MediumLow)

Fig. 11 Rules for fuzzy system 3

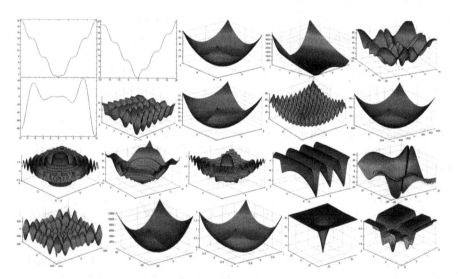

Fig. 12 Benchmark mathematical functions

Fig. 13 Proposal for fuzzy
dynamic adaptation of PSO

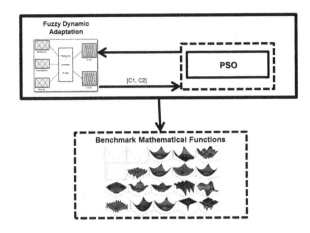

3 Experimentation with the Fuzzy Systems and the Benchmark Mathematical Functions

For the experiments we used the parameters contained in Table 1. Table 1 shows the parameters of the methods to be compared; in this case, we perform a comparison of the proposed method and its variations against the simple PSO algorithm.

Since functions do not have the same global minimum, for comparison it was decided to normalize the results of each function, for this it is used Eq. 8, which gives results between 0 and 1, which means that a number close to 0 is better than a number close to 1.

$$Experiment\ Norm = \left| \frac{Experiment - GlobalMin}{GlobalMax - GlobalMin} \right| \tag{8}$$

To normalize the results with Eq. 8, we need the maximum and the minimum of each benchmark mathematical function; in our case these data are known. Also, the absolute value is needed, because we want to know how much difference between the results of the experiment and the minimum value of the function. Therefore, Table 2 shows some experimental results of each method with each function.

Table 1 Parameters for each method

Parameter	Simple PSO	Fuzzy PSO 1	Fuzzy PSO 2	Fuzzy PSO 3
Population	10	10	10	10
Iterations	30	30	30	30
C1	1	Dynamic	Dynamic	Dynamic
C2	3	Dynamic	Dynamic	Dynamic

Table 2 Simulation results

Function	Minimum	Simple PSO	Fuzzy PSO 1	Fuzzy PSO 2	Fuzzy PSO 3
1	1	0.0005	0.0000	0.0001	0.0003
2	0	0.0000	0.0000	0.0000	0.0000
3	0	0.0000	0.0009	0.0000	0.0000
4	0	0.0000	0.0000	0.0000	0.0000
5	−20	0.1042	0.0665	0.0728	0.0743
6	−100.2238	0.1275	0.1277	0.0000	0.0000
7	−18.5547	0.1929	0.2484	0.2645	0.1253
8	0	0.0000	0.0000	0.0003	0.0000
9	0	0.0017	0.0039	0.0157	0.0019
10	0	0.0000	0.0000	0.0001	0.0000

4 Statistical Comparison

To perform the statistical comparison, we have:

3 methods to compare against the simple PSO, (FPSO1, FPSO2, FPSO3).

27 Benchmark mathematical functions.

10 experiments were performed for each method by each function, so it has a total of 270 experiments for each method. Of this total, we took a random sample of 50 experiments for each method for statistical comparison.

The statistical test used for comparison is the z-test, whose parameters are defined in Table 3.

With the parameters in Table 3, we applied the statistical z-test, giving the following results (Table 4):

In applying the statistic z-test, with significance level of 0.05, and the alternative hypothesis says that the average of the proposed method is lower than the average of simple PSO, and of course the null hypothesis tells us that the average of the proposed method is greater than or equal to the average of simple PSO, with a rejection region for all values fall below −1.645. So the statistical test results are that: for the fuzzy PSO 1, there is significant evidence to reject the null hypothesis, as in the fuzzy PSO 3. But in the fuzzy PSO 2, there is no significant evidence to

Table 3 Parameters for the statistical z-test

Parameter	Value
Level of significance	95 %
Alpha	0.05 %
Ha	$\mu 1 < \mu 2$
H0	$\mu 1 \geq \mu 2$
Critical value	−1.645

Table 4 Results of applying statistical z-test

Our method	Simple method	Z value	Evidence
FPSO1	Simple PSO	−2.1937	Significant
FPSO2	Simple PSO	−0.6801	Not significant
FPSO3	Simple PSO	−2.1159	Significant

reject the null hypothesis. In conclusion, two of the proposed variants of PSO were significantly better than simple PSO.

We proposed a method for dynamic adaptation of the parameters of PSO to improve the quality of results. With the results of the statistic test, we can conclude that there is significant evidence to say that the proposed approach could help in the adaptation of parameters in PSO.

Future work includes experiments with functions with more than two dimensions, comparison with other approaches of PSO, for example, PSO with inertia weight and PSO with constriction. Also try to achieve better results for the PSO with fuzzy system 2, more specifically with the input error and the rules of the fuzzy system. In future work also we try to apply the proposed method to other types of problems, for example, optimization of fuzzy systems.

5 Fuzzy Classifier Design

To design fuzzy classifiers were used methods such as PSO simple and proposed methods with parameters adapted dynamically. These methods were applied to different dataset taken from [1, 2, 3, 5, 8, 12, 13], in which the goal is to obtain a fuzzy classifier that "classify" the data in the best way possible. Figure 14 shows an example of a dataset, in this case the Fisher's Iris dataset [5], which shows that it has 4 attributes (length and width of sepal and petal length and width), 150 records and three distinct classes. The figure shows some graphs which visually compares the attributes of Fisher's Iris dataset.

6 Methodology for Designing Fuzzy Classifiers

The methodology proposed for the design of fuzzy classifiers, defined below:

1. Given a dataset, is obtained the number of different classes, and is divided into 70 % for training and 30 % exclusively for testing.
2. From the training data are obtained some necessary features, such as: number of attributes, attribute ranges, etc.
3. It generates a fuzzy classifier of base, from the characteristics obtained.
4. Optimize the rules from fuzzy classifier using the data for training.

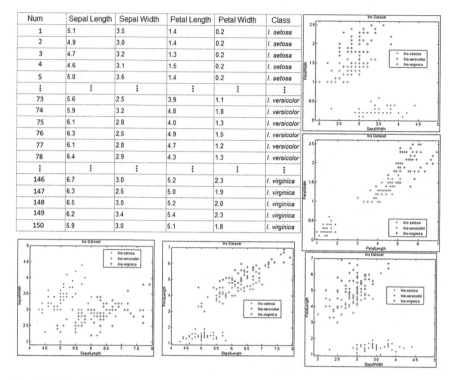

Num	Sepal Length	Sepal Width	Petal Length	Petal Width	Class
1	5.1	3.5	1.4	0.2	I. setosa
2	4.9	3.0	1.4	0.2	I. setosa
3	4.7	3.2	1.3	0.2	I. setosa
4	4.6	3.1	1.5	0.2	I. setosa
5	5.0	3.6	1.4	0.2	I. setosa
⋮	⋮	⋮	⋮	⋮	⋮
73	5.6	2.5	3.9	1.1	I. versicolor
74	5.9	3.2	4.8	1.8	I. versicolor
75	6.1	2.8	4.0	1.3	I. versicolor
76	6.3	2.5	4.9	1.5	I. versicolor
77	6.1	2.8	4.7	1.2	I. versicolor
78	6.4	2.9	4.3	1.3	I. versicolor
⋮	⋮	⋮	⋮	⋮	⋮
146	6.7	3.0	5.2	2.3	I. virginica
147	6.3	2.5	5.0	1.9	I. virginica
148	6.5	3.0	5.2	2.0	I. virginica
149	6.2	3.4	5.4	2.3	I. virginica
150	5.9	3.0	5.1	1.8	I. virginica

Fig. 14 Fisher's Iris dataset

5. Test the best fuzzy classifier found, so far, with the test data.
6. Optimize the membership functions of the best fuzzy classifier found, which has optimized rules, using data for training.
7. Test the best fuzzy classifier found with the test data.

The following defines each step of the methodology for the design of fuzzy classifiers.

In the first step, given a dataset, we obtain the number of different classes, then is divide the dataset, is taken 30 % of the dataset randomly without replacement, the remaining records, that is, 70 % becomes our set of data for training, but before that, are obtained the number of different classes of 70 %, and if this is less than the total of different classes of full dataset, we proceed to select a new random 30 % of records, until the 70 % of the data contains at least one record of each class, this to guarantee that the fuzzy classifier have all possible output classes.

In the second step, to obtain the necessary features from the training data, we obtain the number of attributes, and ranges are obtained from the minimum and maximum of each attribute.

In the third step, to generate a fuzzy classifier, a structure is created to define a Sugeno-type fuzzy system from scratch, where the inputs are each attribute of the dataset and the ranges of the inputs are the ranges of each attribute, and the output of the fuzzy classification system, is given by the total number of classes. Also defines the number of membership functions per input, since the system is Sugeno each output, that is each class, will be an integer. The rules of the fuzzy classifier are all possible combinations in the antecedents, and all the consequents are the first class. Figure 15 shows a fuzzy classifier for Fisher's Iris dataset.

In Fig. 15 one can observe the fuzzy system for classification of Iris dataset. The Iris dataset has four attributes, which are: sepal length, sepal width, petal length and petal width, these attributes are reflected in the inputs of the fuzzy system.

Figure 16 shows the inputs and the output of the fuzzy system, as you can see, each entry has its own range, defined by the training data, in addition, each input has two membership functions, placed symmetrically (can be more membership functions but in the example used only two), and has every possible output dataset class.

Figure 17 shows the set of rules of the fuzzy system for classification of the Iris dataset. As can be observed the number of rules is defined by the number of inputs and their membership functions, in the example, there are 4 inputs with 2 membership functions each, so the maximum number of possible combinations in the antecedents is 16 (that is $2 \times 2 \times 2 \times 2$ or 2^4), plus all rules have class number 1 as consequent, that is for simplicity, since the next step is an optimization of these rules.

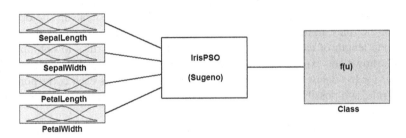

Fig. 15 Fuzzy classifier for Iris dataset

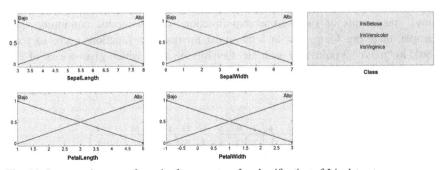

Fig. 16 Inputs and outputs from the fuzzy system for classification of Iris dataset

1. If (SepalLength is Bajo) and (SepalWidth is Bajo) and (PetalLength is Bajo) and (PetalWidth is Bajo) then (Class is IrisSetosa)
2. If (SepalLength is Bajo) and (SepalWidth is Bajo) and (PetalLength is Bajo) and (PetalWidth is Alto) then (Class is IrisSetosa)
3. If (SepalLength is Bajo) and (SepalWidth is Bajo) and (PetalLength is Alto) and (PetalWidth is Bajo) then (Class is IrisSetosa)
4. If (SepalLength is Bajo) and (SepalWidth is Bajo) and (PetalLength is Alto) and (PetalWidth is Alto) then (Class is IrisSetosa)
5. If (SepalLength is Bajo) and (SepalWidth is Alto) and (PetalLength is Bajo) and (PetalWidth is Bajo) then (Class is IrisSetosa)
6. If (SepalLength is Bajo) and (SepalWidth is Alto) and (PetalLength is Bajo) and (PetalWidth is Alto) then (Class is IrisSetosa)
7. If (SepalLength is Bajo) and (SepalWidth is Alto) and (PetalLength is Alto) and (PetalWidth is Bajo) then (Class is IrisSetosa)
8. If (SepalLength is Bajo) and (SepalWidth is Alto) and (PetalLength is Alto) and (PetalWidth is Alto) then (Class is IrisSetosa)
9. If (SepalLength is Alto) and (SepalWidth is Bajo) and (PetalLength is Bajo) and (PetalWidth is Bajo) then (Class is IrisSetosa)
10. If (SepalLength is Alto) and (SepalWidth is Bajo) and (PetalLength is Bajo) and (PetalWidth is Alto) then (Class is IrisSetosa)
11. If (SepalLength is Alto) and (SepalWidth is Bajo) and (PetalLength is Alto) and (PetalWidth is Bajo) then (Class is IrisSetosa)
12. If (SepalLength is Alto) and (SepalWidth is Bajo) and (PetalLength is Alto) and (PetalWidth is Alto) then (Class is IrisSetosa)
13. If (SepalLength is Alto) and (SepalWidth is Alto) and (PetalLength is Bajo) and (PetalWidth is Bajo) then (Class is IrisSetosa)
14. If (SepalLength is Alto) and (SepalWidth is Alto) and (PetalLength is Bajo) and (PetalWidth is Alto) then (Class is IrisSetosa)
15. If (SepalLength is Alto) and (SepalWidth is Alto) and (PetalLength is Alto) and (PetalWidth is Bajo) then (Class is IrisSetosa)
16. If (SepalLength is Alto) and (SepalWidth is Alto) and (PetalLength is Alto) and (PetalWidth is Alto) then (Class is IrisSetosa)

Fig. 17 Set of rules from the fuzzy classifier

Only for purposes of assessing the evolution of the fuzzy classifier, we applied this to the classification of the test data, that to obtain an error of classification using Eq. 9. The error in this case is 66.67 %, this because the consequent of all rules is Class 1, and since the dataset contains 3 classes with 50 records each, this is, the third part of the dataset is for a class.

$$Error\ of\ classification = \frac{Misclassified\ Records}{Total\ Records} \qquad (9)$$

For the fourth step, are used the PSO methods (simple and proposed), for optimization of the rules of the fuzzy classifier previously generated. The optimization of the rules, in this case, is the modification of the consequents of the rules, so that the length of each particle depends on the number of rules to optimize, in Fig. 18 shown an example of a particle for optimizing the rules of the fuzzy classifier for Iris dataset, where you can see it has 16 positions each corresponding to each rule, and possible values are the numbers 1, 2 or 3, which correspond to the possible classes.

For the fifth step, simply is used the best fuzzy classifier found in the previous step, to classify the test data and obtain an error of classification in this case of 22.67 %. As can be seen by comparing errors before optimize rules (66.67 %) and once optimized, there is an improvement in the data classification.

For the sixth step, the optimization of the membership functions consist in "move" the points of the membership functions of the inputs, continuing the example, there are 4 inputs with 2 triangular membership functions each, so you should to "move" 3 points for each membership function, which gives a total of 24 points, these 24 points become the size of the particle. Figure 19 shows an example of a particle for optimizing the membership functions of the fuzzy classifier system.

1	2	3	4	5	6	7	8	9	10	11	12	13	14	15	16
1	3	1	1	1	1	2	3	1	1	3	3	1	2	2	3

Fig. 18 Particle for optimizing the rules of the fuzzy classifier

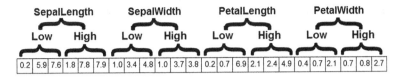

Fig. 19 Particle for optimizing the membership functions of the fuzzy classifier

For the last step uses the best fuzzy classifier found, after optimization of rules and membership functions to classify the test data and obtain a classification error. In this case the classification error is 14 % compared with the classification error optimized keeping only the rules (22.67 %), so that there is an improvement in the classification of data, and is saved the fuzzy classifier with less error.

Figure 20 shown the rules optimized of the fuzzy classifier, and Fig. 21 shown the membership functions of each input once optimized.

1. If (SepalLength is Bajo) and (SepalWidth is Bajo) and (PetalLength is Bajo) and (PetalWidth is Bajo) then (Class is IrisSetosa)
2. If (SepalLength is Bajo) and (SepalWidth is Bajo) and (PetalLength is Bajo) and (PetalWidth is Alto) then (Class is IrisVersicolor)
3. If (SepalLength is Bajo) and (SepalWidth is Bajo) and (PetalLength is Alto) and (PetalWidth is Bajo) then (Class is IrisSetosa)
4. If (SepalLength is Bajo) and (SepalWidth is Bajo) and (PetalLength is Alto) and (PetalWidth is Alto) then (Class is IrisSetosa)
5. If (SepalLength is Bajo) and (SepalWidth is Alto) and (PetalLength is Bajo) and (PetalWidth is Bajo) then (Class is IrisSetosa)
6. If (SepalLength is Bajo) and (SepalWidth is Alto) and (PetalLength is Bajo) and (PetalWidth is Alto) then (Class is IrisVirginica)
7. If (SepalLength is Bajo) and (SepalWidth is Alto) and (PetalLength is Alto) and (PetalWidth is Bajo) then (Class is IrisVirginica)
8. If (SepalLength is Bajo) and (SepalWidth is Alto) and (PetalLength is Alto) and (PetalWidth is Alto) then (Class is IrisVirginica)
9. If (SepalLength is Alto) and (SepalWidth is Bajo) and (PetalLength is Bajo) and (PetalWidth is Bajo) then (Class is IrisSetosa)
10. If (SepalLength is Alto) and (SepalWidth is Bajo) and (PetalLength is Bajo) and (PetalWidth is Alto) then (Class is IrisVersicolor)
11. If (SepalLength is Alto) and (SepalWidth is Bajo) and (PetalLength is Alto) and (PetalWidth is Bajo) then (Class is IrisSetosa)
12. If (SepalLength is Alto) and (SepalWidth is Bajo) and (PetalLength is Alto) and (PetalWidth is Alto) then (Class is IrisVirginica)
13. If (SepalLength is Alto) and (SepalWidth is Alto) and (PetalLength is Bajo) and (PetalWidth is Bajo) then (Class is IrisSetosa)
14. If (SepalLength is Alto) and (SepalWidth is Alto) and (PetalLength is Bajo) and (PetalWidth is Alto) then (Class is IrisVirginica)
15. If (SepalLength is Alto) and (SepalWidth is Alto) and (PetalLength is Alto) and (PetalWidth is Bajo) then (Class is IrisSetosa)
16. If (SepalLength is Alto) and (SepalWidth is Alto) and (PetalLength is Alto) and (PetalWidth is Alto) then (Class is IrisVersicolor)

Fig. 20 Set of optimized rules of the fuzzy classifier

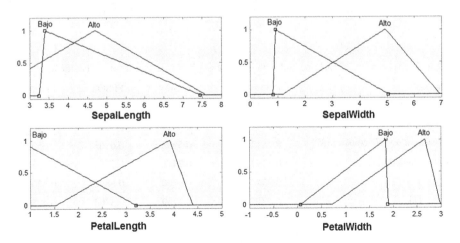

Fig. 21 Optimized membership functions of the fuzzy classifier

7 Experimentation in the Design of Fuzzy Classifiers

For experimentation in the design of fuzzy classifiers, we used dataset taken from [1, 2, 3, 5, 8, 12, 13]. Table 5 shows the main features of these dataset.

As can be seen in Table 5, the data supported by the proposed method (so far) are: numerical and categorical. Addition the proposed method can work with a varied number of attributes (inputs for fuzzy classifier), number of classes (outputs) and number of instances (records to classify).

The parameters used for each method are the same as specified by performing experiments with benchmark mathematical functions, this is, the parameters included in Table 1.

Table 6 shows some of the results of experiments in the design of fuzzy classifiers (first results were taken of each method with each dataset, and in total 10 experiments were performed for each method with each dataset). Moreover, these experiments using Eq. 9, so that have values between the ranges of 0–1.

Table 5 Dataset used in the design of fuzzy classifiers

Name	Instances	Classes	Attributes	Types of data
Abalone [12]	4,177	28	8	Numeric and categorical
Breast tissue [8]	106	6	9	Numeric
Breast cancer Wisconsin [13]	699	2	6	Numeric and categorical
Car evaluation [2]	1,728	4	6	Categorical
Iris [5]	150	3	4	Numeric
Wine [1]	178	3	13	Numeric
Wine quality red [3]	4,898	11	12	Numeric
Wine quality white [3]	4,898	11	12	Numeric

Table 6 Experiments of each method with each dataset in the design of fuzzy classifiers

Dataset	PSO simple	FPSO1	FPSO2	FPSO3
Abalone [12]	0.3556	0.1333	0.2667	0.2222
Breast tissue [8]	0.0667	0.0714	0.1238	0.0905
Breast cancer Wisconsin [13]	0.7813	0.5938	0.6563	0.6563
Car evaluation [2]	0.7399	0.7418	0.7784	0.6956
Iris [5]	0.9346	0.8900	0.8892	0.8708
Wine [1]	0.4815	0.4259	0.3704	0.4259
Wine quality red [3]	0.5208	0.4729	0.4917	0.4417
Wine quality white [3]	0.5687	0.5844	0.5531	0.5401

Table 7 Parameters for statistical z-test

Parameter	Value
Level of significance	95 %
Alpha	0.05 %
Ha	$\mu1 < \mu2$
H0	$\mu1 \geq \mu2$
Critical value	−1.645

8 Statistical Comparison for Fuzzy Classifiers

For the statistical comparison we have:

3 methods to compare against PSO simple, which are: FPSO1, FPSO2, FPSO3 with 8 datasets.

10 experiments were performed for each method with each dataset, so that it has a total of 80 experiments per method. From this total, took a random sample of 30 experiments for each method.

The statistical test used for comparison is the z-test, whose parameters are defined in Table 7.

With the results contained in Table 7, we applied the statistical z-test, obtaining the results contained in Table 8.

In applying the statistic z-test, with significance level of 0.05, and the alternative hypothesis says that the average of the proposed method is lower than the average of simple PSO, and of course the null hypothesis tells us that the average of the proposed method is greater than or equal to the average of simple PSO, with a rejection region for all values fall below −1.645. So the statistical test results are that: for the FPSO1, there is significant evidence to reject the null hypothesis. But in the FPSO2 and FPSO3, there is no significant evidence to reject the null hypothesis.

One of the main reasons that the methods FPSO2 and FPSO3, have not found sufficient statistical evidence to reject the null hypothesis, is because both use the variable input error, and given that this variable needs to know the minimum and maximum of each experiment, this made the use of this variable does not give good results.

In analyzing the results of the statistical test in the design of fuzzy classifiers can see that only the first method found statistical evidence to reject the null hypothesis, i.e., that the proposed method obtains less error in designing fuzzy classifiers.

Table 8 Results of applying statistical z-test

Our method	Simple method	Z value	Evidence
FPSO1	Simple PSO	−2.1502	Significant
FPSO2	Simple PSO	−0.8841	Not significant
FPSO3	Simple PSO	−1.4242	Not significant

The reason that only the first proposed method obtains good results is because when fuzzy systems are designed to adjust parameters, was taken as a premise, that at the beginning the optimization method (PSO here), should explore the search space to explode eventually found the best area, and to do this, the best variables to use are the iteration and diversity, which are precisely the entries of the first method and the other two methods involve variable error, it is for this reason that the first method can handle diversity and convergence in a better way than the other two methods.

9 Conclusions

We conclude that dynamically adjust parameters of an optimization method (in this case the particle swarm optimization PSO), can improve the quality of results and increase the diversity of solutions to a problem.

Three fuzzy systems were designed for adjusting the parameters for particle swarm optimization, which was obtained in two systems statistical evidence of an improvement in the quality of the results of the method of particle swarm optimization when applied in the minimization of benchmark mathematical functions.

Experiments were conducted with the proposed methods in the minimization of mathematical functions and the design of fuzzy classifiers, and a comparison was made between the method of simple particle swarm optimization and the proposed methods, i.e. with fuzzy parameters adjustment.

By comparing the proposed methods and the simple method of PSO, in the design of fuzzy classifiers was found that only the first method obtained statistical evidence to reject the null hypothesis, which says that in developing this thesis was possible to develop a method for adjusting the parameters C1 and C2 of the PSO using fuzzy logic. And in this way improve the results compared with the simple method of PSO.

References

1. Aeberhard, S., Coomans, D., de Vel O.: Comparison of classifiers in high dimensional settings. Technical Report no. 92-02, (1992), Department of Computer Science and Department of Mathematics and Statistics, James Cook University of North Queensland
2. Bohanec, M., Rajkovic, V.: Knowledge acquisition and explanation for multi-attribute decision making. In: 8th International Workshop on Expert Systems and their Applications, pp. 59–78. Avignon, France (1988)
3. Cortez, P., Cerdeira, A., Almeida, F., Matos, T., Reis, J.: Modeling wine preferences by data mining from physicochemical properties. Decis. Support Syst. 47(4), 547–553 (2009)
4. Engelbrecht, A.: Fundamentals of Computational Swarm Intelligence. University of Pretoria, South Africa
5. Fisher, R.: The use of multiple measurements in taxonomic problems. Ann. Eugenics 7, 179–188 (1936)

6. Haupt, R., Haupt, S.: Practical Genetic Algorithms, 2nd edn. A Wiley-Interscience publication, New Jersey (1988)
7. Jang, J., Sun, C., Mizutani, E.: Neuro-fuzzy and soft computing: a computational approach to learning and machine intelligence. Prentice-Hall, Upper Saddle River (1997)
8. Jossinet, J.: Variability of impedivity in normal and pathological breast tissue. Med. Biol. Eng. Comput. **34**, 346–350 (1996)
9. Kennedy, J., Eberhart, R.: Particle swarm optimization. In: Proceedings of IEEE International Conference on Neural Networks, IV, pp. 1942–1948. IEEE Service Center, Piscataway (1995)
10. Kennedy, J., Eberhart, R.: Swarm Intelligence. Morgan Kaufmann, San Francisco (2001)
11. Marcin, M., Smutnicki, C.: Test functions for optimization needs. Available at: http://www.bioinformaticslaboratory.nl/twikidata/pub/Education/NBICResearchSchool/Optimization/VanKampen/BackgroundInformation/TestFunctions-Optimization.pdf (2005)
12. Waugh, S.: Extending and benchmarking cascade-correlation. PhD thesis, Computer Science Department, University of Tasmania (1995)
13. Wolberg, W., Mangasarian, O.: Multisurface method of pattern separation for medical diagnosis applied to breast cytology. Proc. Nat. Acad. Sci. **87**, 9193–9196 (1990)
14. Zadeh, L.: Fuzzy sets. Inf. Control **8**, 338–353 (1965)

Differential Evolution with Dynamic Adaptation of Parameters for the Optimization of Fuzzy Controllers

Patricia Ochoa, Oscar Castillo and José Soria

Abstract The proposal described in this paper uses the Differential Evolution (DE) algorithm as an optimization method in which we want to dynamically adapt its parameters using fuzzy logic control systems, with the goal that the fuzzy system calculates the optimal parameter of the DE algorithm to find better results, depending on the type of problems the DE is applied. In this case we consider a fuzzy system to dynamically change the variable F.

1 Introduction

The use of fuzzy logic in evolutionary computing is becoming a common approach to improve the performance of the algorithms [15, 16, 17]. Currently the parameters involved in the algorithms are determined by trial and error. In this aspect we propose the application of fuzzy logic which is responsible in performing the dynamic adjustment of mutation and crossover parameters in the Differential Evolution (DE) algorithm. This has the goal of providing better performance to Differential Evolution.

Fuzzy logic or multi-valued logic is based on fuzzy set theory proposed by Zadeh in 1965 which helps us in modeling knowledge, through the use of if-then fuzzy rules. The fuzzy set theory provides a systematic calculus to deal with linguistic information, and that improves the numerical computation by using linguistic labels stipulated by membership functions [12]. Differential Evolution (DE) is one of the latest evolutionary algorithms that have been proposed. It was created in 1994 by Price and Storn in, attempts to resolve the problem of Chebychev polynomial. The following year these two authors proposed the DE for optimization of nonlinear and non-differentiable functions on continuous spaces.

P. Ochoa · O. Castillo (✉) · J. Soria
Tijuana Institute of Technology, Tijuana, Mexico
e-mail: ocastillo@tectijuana.mx

© Springer International Publishing Switzerland 2015
O. Castillo and P. Melin (eds.), *Fuzzy Logic Augmentation of Nature-Inspired Optimization Metaheuristics*, Studies in Computational Intelligence 574,
DOI 10.1007/978-3-319-10960-2_3

The DE algorithm is a stochastic method of direct search, which has proven to be effective, efficient and robust in a wide variety of applications such as learning of a neural network, a filter design of IIR, aerodynamically optimized. The DE has a number of important features which make it attractive for solving global optimization problems, among them are the following: it has the ability to handle non-differentiable, nonlinear and multimodal objective functions, usually converges to the optimal uses with few control parameters, etc.

The DE belongs to the class of evolutionary algorithms that is based on populations. It uses two evolutionary mechanisms for the generation of descendants: mutation and crossover; finally a replacement mechanism, which is applied between the vector father and son vector determining who survive into the next generation. There exist works where they currently use fuzzy logic to optimize the performance of the algorithms, to name a few articles such as:

Optimization of Membership Functions for Type-1 and Type 2 Fuzzy Controllers of an Autonomous Mobile Robot Using PSO [1], Optimization of a Fuzzy Tracking Controller for an Autonomous Mobile Robot under Perturbed Torques by Means of a Chemical Optimization Paradigm [2], Design of Fuzzy Control Systems with Different PSO Variants [4], A Method to Solve the Traveling Salesman Problem Using Ant Colony Optimization Variants with Ant Set Partitioning [6], Evolutionary Optimization of the Fuzzy Integrator in a Navigation System for a Mobile Robot [7], Optimal design of fuzzy classification systems using PSO with dynamic parameter adaptation through fuzzy logic [8], : Dynamic Fuzzy Logic Parameter Tuning for ACO and Its Application in TSP Problems [10], Bio-inspired Optimization Methods on Graphic Processing Unit for Minimization of Complex Mathematical Functions [18].

Similarly there are papers on Differential Evolution (DE) applications that uses this algorithm to solve real problems. To mention a few:

A fuzzy logic control using a differential evolution algorithm aimed at modelling the financial market dynamics [5], Design of optimized cascade fuzzy controller based on differential evolution: Simulation studies and practical insights [11], Eliciting transparent fuzzy model using differential evolution [3], Assessment of human operator functional state using a novel differential evolution optimization based adaptive fuzzy model [14, 20].

This paper is organized as follows: Sect. 2 shows the concept of the Differential Evolution algorithm. Section 3 describes the proposed methods. Section 4 the Benchmark Functions, Sect. 5 the proposed Fuzzy System, Sect. 6 Experiments and Methodology, Sect. 7 shows the Simulation Results and Sect. 8 the Conclusions.

2 Differential Evolution

The Differential Evolution (DE) is an optimization method belonging to the category of evolutionary computation applied in solving complex optimization problems.

The DE is composed of 4 steps:

Initialization.
Mutation.
Crossover.
Selection.

This is a non-deterministic technique based on the evolution of a vector population (individuals) of real values representing the solutions in the search space. The generation of new individuals is carried out by differential crossover and mutation operators [13].

The operation of the algorithm is explained below.

2.1 Population Structure

The differential evolution algorithm maintains a pair of vector populations, both of which contain Np D-dimensional vectors of real-valued parameters [8].

$$P_{x,g} = (x_{i,g}), i = 0, 1, \ldots, Np, g = 0, 1, \ldots, g_{max} \tag{1}$$

$$x_{i,g} = (x_{j,i,g}), j = 0, 1, \ldots, D - 1 \tag{2}$$

where
P_x current population
g_{max} maximum number of iterations
i   index population
j   parameters within the vector

Once the vectors are initialized, three individuals are selected randomly to produce an intermediate population, $P_{v,g}$, of Np mutant vectors, $v_{i,g}$.

$$P_{v,g} = (v_{i,g}), \ i = 0, 1, \ldots, Np - 1, \ g = 0, 1, \ldots, g_{max} \tag{3}$$

$$v_{i,g} = (v_{j,I,g}), \ j = 0, 1, \ldots, D - 1 \tag{4}$$

Each vector in the current population are recombined with a mutant vector to produce a trial population, P_u, the NP, mutant vector $u_{i,g}$:

$$P_{v,g} = (u_{i,g}), \ i = 0, 1, \ldots, Np - 1, \ g = 0, 1, \ldots, g_{max} \tag{5}$$

$$u_{i,g} = (u_{j,I,g}), \ j = 0, 1, \ldots, D - 1 \tag{6}$$

2.2 Initialization

Before initializing the population, the upper and lower limits for each parameter must be specified. These 2D values can be collected by two initialized vectors, D-dimensional, b_L y b_U, to which subscripts L and U indicate the lower and upper limits respectively. Once the initialization limits have been specified number generator randomly assigns each parameter in every vector a value within the set range. For example, the initial value (g = 0) of the j-th vector parameter is i-th:

$$x_{j,i,0} = rand_j(0,1) \cdot (b_{j,U} - b_{j,L}) + b_{j,L} \tag{7}$$

2.3 Mutation

In particular, the differential mutation adds a random sample equation showing how to combine three different vectors chosen randomly to create a mutant vector.

$$v_{i,g} = \mathbf{x}_{r0,g} + F \cdot (\mathbf{x}_{r1,g} - \mathbf{x}_{r2,g}) \tag{8}$$

The scale factor, $F \in (0, 1)$ is a positive real number that controls the rate at which the population evolves. While there is no upper limit on F, the values are rarely greater than 1.0.

2.4 Crossover

To complement the differential mutation search strategy, DE also uses uniform crossover. Sometimes known as discrete recombination (dual). In particular, DE crosses each vector with a mutant vector:

$$U_{i,g} = (u_{j,i,g}) = \begin{cases} v_{j,i,g} & \text{if } (rand_j(0,1) \le Cr \text{ or } j = j_{rand}) \\ x_{j,i,g} & \text{otherwise.} \end{cases} \tag{9}$$

2.5 Selection

If the test vector, $U_{i,g}$ has a value of the objective function equal to or less than its target vector, $\mathbf{X}_{i,g}$. It replaces the target vector in the next generation; otherwise, the target retains its place in population for at least another generation [2]

$$X_{i,g+1} = \begin{cases} U_{i,g} & \text{if } f(U_{i,g}) \le f(X_{i,g}) \\ X_{i,g} & \text{otherwise.} \end{cases} \tag{10}$$

The process of mutation, recombination and selection are repeated until the optimum is found, or terminating pre criteria specified is satisfied. DE is a simple, but powerful search engine that simulates natural evolution combined with a mechanism to generate multiple search directions based on the distribution of solutions in the current population. Each vector i in the population at generation G, xi,G, called at this moment of reproduction as the target vector will be able to generate one offspring, called trial vector (ui,G). This trial vector is generated as follows: First of all, a search direction is defined by calculating the difference between a pair of vectors $r1$ and $r2$, called "*differential vectors*", both of them chosen at random from the population. This difference vector is also scaled by using a user defined parameter called "$F \geq 0$". This scaled difference vector is then added to a third vector $r3$, called "*base vector*". As a result, a new vector is obtained, known as the mutation vector. After that, this mutation vector is recombined with the target vector (also called parent vector) by using discrete recombination (usually binomial crossover) controlled by a crossover parameter $0 \leq CR \leq 1$ whose value determines how similar the trial vector will be with respect to the target vector. There are several DE variants. However, the most known and used is DE/rand/1/ bin, where the base vector is chosen at random, there is only a pair of differential vectors and a binomial crossover is used. The detailed pseudocode of this variant is presented in Fig. 1 [9].

Fig. 1 "DE/rand/1/bin" pseudocode rand [0, 1) is a function that returns a real number between 0 and 1. Randint (min, max) is a function that returns an integer number between min and max. *NP, MAX GEN, CR* and *F* are user-defined parameters *n* is the dimensionality of the problem [9]

```
Begin
   G=0
   Create a random initial population x_{i,G} ∀i, i = 1,...,NP
   Evaluate f(x_{i,G}) ∀i, i = 1,...,NP
   For G=1 to MAX_GEN Do
       For i=1 to NP Do
           Select randomly r_1 ≠ r_2 ≠ r_3 :
           j_rand = randint(1,D)
           For j=1 to n Do
               If (rand_j[0,1) < CR or j = j_rand) Then
                   u_{i,j,G+1} = x_{r3,j,G} + F(x_{r1,j,G} − x_{r2,j,G})
               Else
                   u_{i,j,G+1} = x_{i,j,G}
               End If
           End For
           If (f(u_{i,G+1}) ≤ f(x_{i,G})) Then
               x_{i,G+1} = u_{i,G+1}
           Else
               x_{i,G+1} = x_{i,G}
           End If
       End For
       G = G+1
   End For
End
```

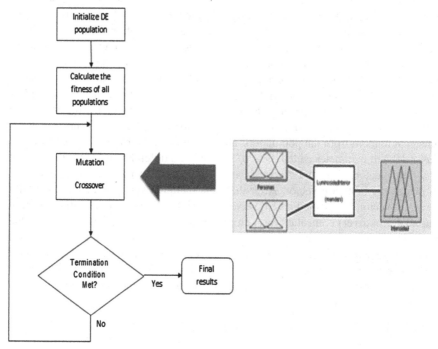

Fig. 2 The proposed differential evolution (DE) algorithm by integrating a fuzzy system to dynamically adapt parameters

3 Proposed Method

The Differential Evolution (DE) Algorithm is a powerful search technique used for solving optimization problems. In this paper a new algorithm called Fuzzy Differential Evolution (FDE) with dynamic adjustment of parameters for the optimization of controllers is proposed. The main objective is that the fuzzy system will provides us with the optimal parameters for the best performance of the DE algorithm. In addition the parameters that the fuzzy system optimizes are the crossover and mutation, as shown in Fig. 2.

4 Benchmark Function

In this paper we consider 6 Benchmark functions which are briefly explained below [19].

Fig. 3 Sphere function

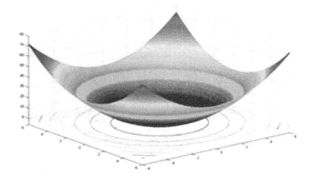

- **Sphere Function**

  Number of variables: n variables.
  Definition:

  $$f(x) = \sum_{i=1}^{n} x_i^2 \tag{11}$$

  Search domain: $-5.12 \leq x_i \leq 5.12$, $i = 1, 2, \ldots, n$.
  Number of local minima: no local minimum except the global one.
  The global minima: $x^* = (0, \ldots, 0)$, $f(x^*) = 0$
  Function graph: for n = 2 presented in Fig. 3

- **Griewank Function**

  Number of variables: n variables.
  Definition:

  $$f(x) = \frac{1}{4000} \sum_{i=1}^{n} x_i^2 - \prod_{i=1}^{n} \cos\left(\frac{x_i}{\sqrt{i}}\right) + 1 \tag{12}$$

  Search domain: $-600 \leq x_i \leq 600$, $i = 1, 2, \ldots, n$.
  Number of local minima: no local minimum except the global one.
  The global minima: $x^* = (0, \ldots, 0)$, $f(x^*) = 0$
  Function graph: for n = 2 presented in Fig. 4

- **Schwefel Function**

  Number of variables: n variables.
  Definition:

  $$f(x) = \sum_{i=1}^{n} \left[-x_i \sin(\sqrt{|x_i|}) \right] \tag{13}$$

Fig. 1 Griewank function

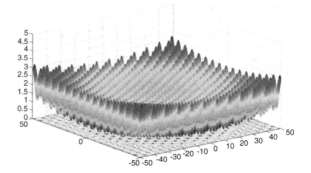

Search domain: $-500 \le x_i \le 500$, $i = 1, 2, \ldots, n$.
Number of local minima: no local minimum except the global one.
The global minima: $x^* = (0, \ldots, 0)$, $f(x^*) = 0$
Function graph: for n = 2 presented in Fig. 5

- **Rastringin Function**

  Number of variables: n variables.
  Definition:

$$f(x) = 10n + \sum_{i=1}^{n} \left[x_i^2 - 10\cos(2\pi x_i) \right] \tag{14}$$

Search domain: $-5.12 \le x_i \le 5.12$, $i = 1, 2, \ldots, n$.
Number of local minima: no local minimum except the global one.
The global minima: $x^* = (0, \ldots, 0)$, $f(x^*) = 0$
Function graph: for n = 2 presented in Fig. 6

- **Ackley Funcion**

  Number of variables: n variables.

Fig. 5 Schwefel function

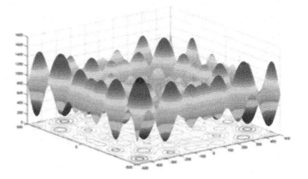

Fig. 6 Rastrigin function

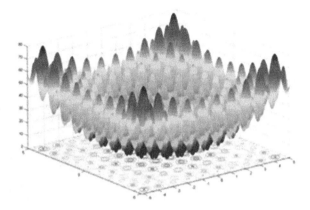

Definition:

$$f(x) = -a \cdot \exp(-b \cdot \sqrt{\frac{1}{n} \sum_{i=1}^{n} x_i^2}) - \exp\left(\frac{1}{n} \sum_{i=1}^{n} \cos(cx_i)\right) + a + \exp(1) \quad (15)$$

Search domain: $-15 \leq x_i \leq 30$, $i = 1, 2, ..., n$.
Number of local minima: no local minimum except the global one.
The global minima: $x^* = (0, ..., 0)$, $f(x^*) = 0$
Function graph: for $n = 2$ presented in Fig. 7

- **Rosenbrock Funcion**

Number of variables: n variables.
Definition:

$$f(x) = \sum_{i=1}^{n-1} \left[100(x_{i+1} - x_i^2)^2 + (1 - x_i)^2 \right]. \quad (16)$$

Fig. 7 Ackley function

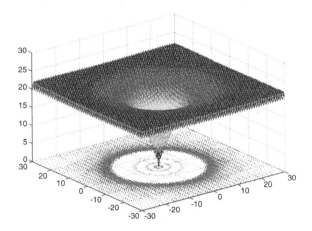

Fig. 8 Rosenbrock function

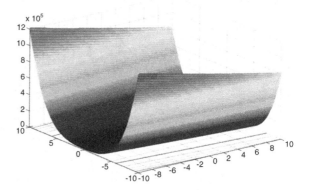

Search domain: $-5 \leq xi \leq 10$, $i = 1, 2, ..., n$.
Number of local minima: no local minimum except the global one.
The global minima: $x^* = (0, ..., 0)$, $f(x^*) = 0$
Function graph: for $n = 2$ presented in Fig. 8

5 Fuzzy System

This paper mentioned two fuzzy systems with which the experiments were performed. It has a fuzzy system which increase the F variable and another variable decrease F.

Then the fuzzy system, in which F is increased dynamically.

- Contains one input and one output
- Is of Mamdani type.
- All membership functions are triangular.
- The input of the fuzzy system is defined by the generations and granulated into three membership functions they are: MF1 = 'Low'[−0.5 0 0.5], MF2 = 'Medium'[0 0.5 1], MF3 = 'High'[0.5 1 1.5].
- The output of the fuzzy system and the variable F is granulated in three membership functions which are: MF1 = 'Low', [−0.5 0 0.5], MF2 = 'Medium', [0 0.5 1] MF3 = 'High', [0.5 1 1.5].
- The fuzzy system uses 3 rules and what it does is increased the value of the F variable in a range of (0.1).

Shown in Fig. 9.

Then the fuzzy system, in which F is dynamically decreased is described as follows:

- Contains one input and one output
- Is Mamdani type.

> 1. - If (Generations is Low) then (F is Low) (1)
> 2. - If (Generations isMedium) then (F is Medium) (1)
> 3. - If (Generations is High) then (F isHigh) (1)

Fig. 9 Rules of the fuzzy system

> 1. - If (Generations is Low) then (F is High) (1)
> 2. - If (Generations is Medium) then (F is Medium) (1)
> 3. - If (Generations is High) then (F isLow) (1)

Fig. 10 Rules of the fuzzy system

- All functions are triangular.
- The input of the fuzzy system is generations and divided into three membership functions they are: MF1 = 'Low'[−0.5 0 0.5], MF2 = 'Medium' [0 0.5 1], MF3 = 'High'[0.5 1 1.5].
- The output of the fuzzy system and the variable F is divided in three membership functions which are: MF1 = 'Low', [−0.5 0 0.5], MF2 = 'Medium', [0 0.5 1] MF3 = 'High', [0.5 1 1.5].
- The fuzzy system uses 3 rules and what it does is decreased the value of the F variable in a range of (0.1).

Shown in Fig. 10.

6 Experiments and Methodology

Experiments with the Differential Evolution algorithm varying the F value (variable mutation) manually in a range from 0.1 to 0.9, were performed with several generations values (GR) 100, 500, 1000, 2000, 3000, 4000 and 5000. Population variables are held constant (NP) = 250, dimension (D) = 50, crossover (CR) = 0.1, upper limit (H) = 500 and lower (L). = -500.

To make a comparison with the differential evolution algorithm the averages of the experiments are obtained for each generation number. Experiments are performed by varying F from 0.1 to 0.9 and 30 experiments for each F. then an overall average is obtained for comparison with the Fuzzy Differential Evolution with increase and decrease of the F value.

For experiments with the Fuzzy Differential Evolution algorithm F changes dynamically increasing and decreasing F between 0 and 1, 30 experiments for each number of generations are performed and an average is obtained by generation.

For functions Ackley and Rosenbrock search space since we were using was very spacious for the Ackley function, the upper limit (H) = 32.768 and the lower limit (L) = −32.768, for the Rosenbrock function the upper limit (H) = 2.048 and the lower limit (L) = −2.048.

7 Simulation Results

In this paper we show five tables to compare the results of the Benchmark functions mentioned above, where the variable F is modified manually in the Difference Evolution algorithm, F is increased and described dynamically with the proposed Fuzzy Differential Evolution algorithm.

Table 1 shows a comparison for the Sphere function, F is manually varied in the Differential Evolution algorithm and in the Fuzzy Differential Evolution algorithm F is increased and decreased dynamically with the number of generations with which the experiments were performed.

Table 2 shows a comparison of the Griewarnk function, F is manually varied in the Differential Evolution algorithm and in the Fuzzy Differential Evolution algorithm F is increased and decreased dynamically with the number of generations with which the experiments were performed.

Table 3 shows a comparison of the Schwefel function, F is manually varied in the Differential Evolution algorithm and in the Fuzzy Differential Evolution algorithm F is increased and decreased dynamically with the number of generations with which the experiments were performed.

Table 4 shows a comparison of the Rastringin function, F is manually varied in the Differential Evolution algorithm and in the Fuzzy Differential Evolution algorithm F is increased and decreased dynamically with the number of generations with which the experiments were performed.

Table 5 shows a comparison of the Ackley function, F is manually varied in the Differential Evolution algorithm and in the Fuzzy Differential Evolution algorithm F is increased and decreased dynamically with the number of generations with which the experiments were performed.

Table 6 shows a comparison of the Rosenbrock function, F is manually varied in the Differential Evolution algorithm and in the Fuzzy Differential Evolution algorithm F is increased and decreased dynamically with the number of generations with which the experiments were performed.

Table 1 Sphere function

Generations	Differential evolution	Fuzzy differential evolution with increasing F	Fuzzy differential evolution with decreasing F
100	274810.954	264876.554	191810.002
500	3416.56935	35.9248183	19.9166364
1000	3.91E+01	0.00048186	0.00024121
2000	9.61E−03	7.2858E−14	3.6657E−14
3000	2.39E−06	1.3168E−23	5.5608E−24
4000	6.10E−10	1.8705E−33	9.4497E−34
5000	1.54E−13	3.1962E−43	1.36E−43

Table 2 Griewarnk function

Generations	Differential evolution	Fuzzy differential evolution with increasing F	Fuzzy differential evolution with decreasing F
100	69.9332605	67.0770188	48.1462646
500	1.37127437	0.6940305	0.51379811
1000	2.07E−01	3.4223E−05	1.836E−05
2000	0.00087914	5.2217E−15	2.2797E−15
3000	2.2279E−07	0	0
4000	5.78723E−11	0	0
5000	1.52742E−14	0	0

Table 3 Schwefel function

Generations	Differential evolution	Fuzzy differential evolution with increasing F	Fuzzy differential evolution with decreasing F
100	10421.718	11169.3347	10932.22075
500	8100.8323	5193.63165	4842.340175
1000	6388.8028	14.5133895	3.958285778
2000	4043.1945	0.00063638	0.000636378
3000	548.84107	0.00063638	0.000636378
4000	0.1285108	0.00063638	0.000636378
5000	0.1323711	0.00063638	0.000636378

Table 4 Rastringin function

Generations	Differential evolution	Fuzzy differential evolution with increasing F	Fuzzy differential evolution with decreasing F
100	277759.787	268207.392	184622.012
500	3774.94974	379.027991	342.483337
1000	2.25E+02	144.499278	150.754739
2000	63.1322564	76.3943085	84.6821796
3000	43.6992981	47.4723588	57.9202967
4000	31.1808589	24.2475708	37.6116134
5000	16.6249025	1.2237E−05	1.1348E−04

Table 5 Ackley function

Generations	Differential evolution	Fuzzy differential evolution with increasing F	Fuzzy differential evolution with decreasing F
100	13.5629222	14.568239	13.3167462
500	1.89192803	0.4362296	0.26047838
1000	2.77E−01	0.00084351	0.00059808
2000	1.43E−03	1.0485E−08	7.5391E−09
3000	2.07E−05	1.2961E−13	1.02E−13
4000	3.27E−07	8.941E−15	7.52E−15
5000	5.11E−09	7.99E−15	6.81E−15

Table 6 Rosenbrock function

Generations	Differential evolution	Fuzzy differential evolution with increasing F	Fuzzy differential evolution with decreasing F
100	193.195633	165.930737	141.722514
500	3.91E+01	44.4584465	34.4383295
1000	1.46E+01	0.30151117	1.54954836
2000	0.05458507	4.3226E−12	2.2078E−10
3000	9.6435E−06	7.95E−23	2.0198E−20
4000	1.7784E−09	0	0
5000	3.4709E−13	0	0

8 Conclusions

We can conclude that the Differential Evolution algorithm with Fuzzy F Decrease performs better than the Differential Evolution algorithm with F increase because in most mathematical functions when a Decrease the EDF algorithm obtained better results. We can also conclude with experiments that manually using a small F we can get better results.

The number of iterations used in the algorithm development differential is greater than the number of iterations used in the Differential Evolution in Fuzzy algorithms in increase or decrease.

Another important aspect to consider for the experiments is the search space that is given to each function, as happened with Ackley and Rosenbrock functions which the search space is modified to obtain better results.

References

1. Aguas-Marmolejo, S.J., Castillo, O.: optimization of membership functions for type-1 and type 2 fuzzy controllers of an autonomous mobile robot using PSO. In: Recent Advances on Hybrid Intelligent Systems, pp. 97–104. Springer, Berlin (2013)
2. Astudillo, L., Melin, P., Castillo, O.: optimization of a fuzzy tracking controller for an autonomous mobile robot under perturbed torques by means of a chemical optimization paradigm. In: Recent Advances on Hybrid Intelligent Systems, pp. 3–20. Springer, Berlin (2013)
3. Eftekhari, M., Katebi, S.D., Karimi, M., Jahanmir, A.H.: Eliciting transparent fuzzy model using differential evolution. Appl. Soft Comput. **8**, 466–476 (2008). (School of Engineering, Shiraz University, Shiraz, Iran)
4. Fierro, R., Castillo, O.: Design of fuzzy control systems with different PSO variants. In: Recent Advances on Hybrid Intelligent Systems, pp. 81–88. Springer, Berlin (2013)
5. Hachicha, N., Jarboui, B., Siarry, P.: A fuzzy logic control using a differential evolution algorithm aimed at modelling the financial market dynamics. Inf. Sci. **181**, 79–91 (2011). (Institut Supérieur de Commerce et de Comptabilité de Bizerte, Zarzouna 7021, Bizerte, Tunisia)

6. Lizárraga, E., Castillo, O., Soria, J.: A method to solve the traveling salesman problem using ant colony optimization variants with ant set partitioning. In: Recent Advances on Hybrid Intelligent Systems, pp. 237–2461. Springer, Berlin (2013)
7. Melendez, A., Castillo, O.: Evolutionary optimization of the fuzzy integrator in a navigation system for a mobile robot. In: Recent Advances on Hybrid Intelligent Systems, pp. 21–31. Springer, Berlin (2013)
8. Melin, P., Olivas, F., Castillo, O., Valdez, F., Soria, J., García, J.: Optimal design of fuzzy classification systems using PSO with dynamic parameter adaptation through fuzzy logic. Expert Syst. Appl. **40**(8), 3196–3206 (2013)
9. Mezura-Montes, E., Palomeque-Ortiz, A.: self-adaptive and deterministic parameter control in differential evolution for constrained optimization. Efren Mezura-Montes, Laboratorio Nacional de Informatica Avanzada (LANIA A.C.), Rebsamen 80, Centro, Xalapa, Veracruz, 91000, MEXICO 2009
10. Neyoy, H., Castillo, O., Soria, J.: dynamic fuzzy logic parameter tuning for ACO and its application in TSP problems. In: Recent Advances on Hybrid Intelligent Systems, pp. 259–271. Springer, Berlin (2013)
11. Oh, S.-K., Kim, W.-D., Pedrycz, W.: Design of optimized cascade fuzzy controller based on differential evolution: Simulation studies and practical insights. Eng. Appl. Artif. Intell. **25**, 520–532 (2012). (Department of Electrical Engineering, The University of Suwon)
12. Olivas, F., Castillo, O.: Particle swarm optimization with dynamic parameter adaptation using fuzzy logic for benchmark mathematical functions. In: Recent Advances on Hybrid Intelligent Systems, pp. 247–258. Springer, Berlin (2013)
13. Price, K.V., Storn, R., Lampinen, J.A.: Differential Evolution. Springer, Berlin (2005)
14. Raofen, W., Zhang, J., Zhang, Y., Wang, X.: Assessment of human operator functional state using a novel differential evolution optimization based adaptive fuzzy model. Biomed. Signal Process. Control **7**, 490–498 (2012). (Lab for Brain-Computer Interfaces and Control, East China University of Science and Technology, Shanghai 200237, Peoples Republic of China)
15. Sombra, A., Valdez, F., Melin, P., Castillo, O.: A new gravitational search algorithm using fuzzy logic to parameter adaptation. In: IEEE Congress on Evolutionary Computation, pp. 1068–1074 (2013
16. Valdez, F., Melin, P., Castillo, O.: Evolutionary method combining particle swarm optimization and genetic algorithms using fuzzy logic for decision making. In: Proceedings of the IEEE International Conference on Fuzzy Systems, pp. 2114–2119 (2009)
17. Valdez, F., Melin, P., Castillo, O.: Parallel particle swarm optimization with parameters adaptation using fuzzy logic. MICAI (2), 374–385 (2012)
18. Valdez, F., Melin, P., Castillo, O.: Bio-inspired optimization methods on graphic processing unit for minimization of complex mathematical functions. In: Recent Advances on Hybrid Intelligent Systems, pp 313–322. Springer, Berlin (2013)
19. Valdez, F., Melin, P., Castillo, O.: An improved evolutionary method with fuzzy logic for combining particle swarm optimization and genetic algorithms. Appl. Soft Comput. **11**, 2625–2632 (2011)
20. Vucetic, D.: Fuzzy differential evolution algorithm. The University of Western Ontario, London (2012)

A New Bat Algorithm with Fuzzy Logic for Dynamical Parameter Adaptation and Its Applicability to Fuzzy Control Design

Jonathan Pérez, Fevrier Valdez and Oscar Castillo

Abstract We describe in this paper the Bat Algorithm and a new approach is proposed using a fuzzy system to dynamically adapt its parameters. The original method is compared with the proposed method and also compared with genetic algorithms, providing a more complete analysis of the effectiveness of the bat algorithm. Simulation results on a set of mathematical functions with the fuzzy bat algorithm outperform the traditional bat algorithm and genetic algorithms and proposed to implement the method in a controller to analyze the effectiveness of the algorithm.

Keywords Bat algorithm · Genetic algorithm · Fuzzy system

1 Introduction

This paper is focus on the study of the Bat Algorithm, which has proven to be one of the best to face problems of nonlinear global optimization.

The bat algorithm is a metaheuristic optimization method proposed by Yang in 2010 and this algorithm is based on the behavior of micro bats echolocation pulses with different emission and sound.

The bat algorithm has the characteristic of being one of the best methods to solve problems of nonlinear global optimization. In this paper the use of the bat algorithm with a fuzzy system is presented with the aim of dynamically setting some of the parameters in the algorithm. The goal is improving the performance of the algorithm against other metaheuristics in optimization problems to validate our approach we used on a set of benchmark mathematical functions.

Once the modification in the bat algorithm is performed, tests were performed with benchmark mathematical functions to analyze its effectiveness. Also the

J. Pérez · F. Valdez (✉) · O. Castillo
Tijuana Institute of Technology, Tijuana, Mexico
e-mail: fevrier@tectijuana.mx

© Springer International Publishing Switzerland 2015
O. Castillo and P. Melin (eds.), *Fuzzy Logic Augmentation of Nature-Inspired Optimization Metaheuristics*, Studies in Computational Intelligence 574,
DOI 10.1007/978-3-319-10960-2_4

original method is compared with the proposed method and genetic algorithms, providing a more complete analysis of the effectiveness of bat algorithm. Simulation results with the fuzzy bat algorithm outperform the traditional bat algorithm and genetic algorithms.

In the current literature there are papers where the bat algorithm has been used, like in the paper A New Metaheuristic Bat-Inspired Algorithm [22]. In this paper, we propose a new metaheuristic method, the Bat Algorithm, making comparison with the PSO and GA algorithms by applying them to Benchmark functions, in the paper the Performance of Firefly and Bat Algorithm for Unconstrained Optimization Problems is shown [6]. In this paper we compare the algorithm against firefly algorithm using bat Benchmark functions, the paper proposes the Bat Algorithm: Literature Review and Applications [23], perform detailed explanation of the bat algorithm and the applications and variants existing at present. In the paper A Comparison of BA, GA, PSO, BP and LM for Training Feed forward Neural Networks in e-Learning Context [10], a comparison of algorithms for training feed forward neural networks was done. Several tests were made on two gradient descent algorithms: Backpropagation and Levenberg-Marquardt, and three population based heuristic: Bat Algorithm, Genetic Algorithm, and Particle Swarm Optimization. Experimental results show that the bat algorithm (BA) outperforms all other algorithms in training feed forward neural networks [10], A Binary Bat Algorithm for Feature Selection [15], performed applying the bat algorithm for feature selection using a binary version of the algorithm, in the paperLocal Memory Search Bat Algorithm for Grey Economic Dynamic System [25], in this paper, the LMSBA is introduced in economic control field, test and simulation results are ideal, and programming of method is concise. This algorithm is suitable for numerical solution in practical dynamic economic control, providing numerical theoretical foundation for steady, healthy and optimal economic growth [15]. In the paper Solving Multi-Stage MultiMachine Multi-Product Scheduling Problem Using Bat Algorithm, the algorithm takes into account the just in time production philosophy by aiming to minimise the combination of earliness and tardiness penalty costs [14]. In the paper use of Fuzzy Systems and Bat Algorithm for Energy Modeling in a Gas Turbine Generator (GTG), is proposed the purpose of this paper has been to demonstrate the use of fuzzy methods to capture variation of exergy destruction in a GTG [1].

In other works listed below on the use of bat algorithm we have, in the article Chaotic bat algorithm [5], performed in which the aggregation of chaos in the standard version of bat algorithm. In the paper A bat-inspired algorithm for structural optimization [8], performing the comparison of several algorithms showing affection bat algorithm in resolving problems of bar among others. In the paper Bat inspired algorithm for discrete size optimization of steel frames [7], is shows, the objective of this study is to investigate efficiency of the bat algorithm in discrete sizing optimization problems of steel frames [7]. In the paper a New Metaheuristic Bat Inspired Classification Approach for Microarray Data [13], the bat algorithm successfully formulated and is used to update the weight of the FLANN classifier. In the paper A wrapper approach for feature selection based on Bat

Algorithm and Optimum-Path Forest [17], is described combines an exploration of the search space and an intense local analysis by exploiting the neighborhood of a good solution to reduce the feature space dimensionality [17], in the paper Bat algorithm for the fuel arrangement optimization of reactor core [9], for the first time, the bat optimization algorithm is applied for the LPO problem. Prior to perform the LPO, the developed BA was validated against a test function obtaining the exact minimum value during various iterations.

This paper is organized as follows in Sect. 2 describe the original bat algorithm, in Sect. 3 describe genetic algorithm, in Sect. 4 describe of the benchmark mathematical functions, in Sect. 5 describe the results between genetic algorithm and bat algorithm, in Sect. 6 describe proposed method and results, in Sect. 7 describe proposed bat algorithm apply a controller and Sect. 8 describe the conclusions.

2 Bat Algorithm

This section describes the basic concepts of the Bat Algorithm.

2.1 Rules of Bats

If we idealize some of the echolocation characteristics of microbats, we can develop various bat-inspired algorithms or bat algorithms. For simplicity, we now use the following approximate or idealized rules [22]:

1. All bats use echolocation to sense distance, ant they also 'know' the difference between food/prey and background barriers in some magical way.
2. Bats fly randomly witch velocity v_i at position x_i with a fixed frequency f_{min}, varying wavelength λ and loudness A_0 to search for prey. They can automatically adjust the wavelength (or frequency) of their emitted pulses and adjust the rate of pulse emission r ∈ [0, 1], depending on the proximity of their target.
3. Although loudness can vary in many ways, we assume that the loudness varies from a large (positive) A_0 to a minimum constant value A_{min}.

For simplicity, the frequency $f \in [0, f_{max}]$, the new solutions x_i^t and velocity v_i^t at a specific time step t are represented by a random vector drawn from a uniform distribution [6].

2.2 Pseudocode for the Bat Algorithm

The basic steps of the bat algorithm, can be summarized as the pseudo code shown in Fig. 1

> *Initialize the bat population $x_i(i=1, 2,..., n)$ and v_i*
> *Initialize frequency f_i, pulse rates r_i and the loudness A_i*
> **While** *(t<Max numbers of iterations)*
>   *Generate new solutions by adjusting frequency*
>   *and updating velocities and locations/solutions [equations (1) to (3)]*
>   *if(rand>r_i)*
>   *Select a solution among the best solutions*
>   *Generate a local solution around the selected best solution*
>   **end if**
>   *Generate a new solutions by flying randomly*
>   *if (rand <A_i& $f(x_i) < f(x_*)$)*
>   *Accept the new solutions*
>   *Increase r_i and reduce A_i*
>   **end if**
>   *Rank the bats and find the current best x_**
> **end while**

Fig. 1 Pseudo code of the bat algorithm

2.3 Movements in the Bat Algorithm

Each bat is associated with a velocity v_i^t and location x_i^t, at iteration t, in a dimensional search or solution space. Among all the bats, there exist a current best solution x_*. Therefore, the above three rules can be translated into the updating equations for x_i^t and velocities v_i^t:

$$f_i = f_{min} + (f_{max} - f_{min})\beta, \tag{1}$$

$$v_i^t = v_i^{t-1} + (x_i^{t-1} - x_*)f_i, \tag{2}$$

$$x_i^t = x_i^{t-1} + v_i^t, \tag{3}$$

where $\beta \in [0, 1]$ is a random vector selected from a uniform distribution [17].

As mentioned earlier, we can either use wavelengths or frequencies for implementation, we will use $f_{min} = 0$ and $f_{max} = 1$, depending on the domain size of the problem of interest. Initially, each bat is randomly assigned a frequency which is drawn uniformly from $[f_{min} - f_{max}]$. The loudness and pulse emission rates essentially provide a mechanism for automatic control and auto zooming into the region with promising solutions [23].

2.4 Loudness and Pulse Rates

In order to provide an effective mechanism to control the exploration and exploitation and switch to exploitation stage when necessary, we have to vary the loudness A_i and the rate r_i of pulse emission during the iterations. Since the loudness usually decreases once a bat has found its prey, while the rate of pulse emission increases, the loudness can be chosen as any value of convenience, between A_{min} and A_{max}, assuming $A_{min} = 0$ means that a bat has just found the prey and temporarily stop emitting any sound. With these assumptions, we have

$$A_i^{t+1} = \alpha A_i^t, r_i^{t+1} = r_i^0[1 - \exp(-\gamma^t)], \tag{4}$$

where α and γ are constants. In essence, here α is similar to the cooling factor of a cooling schedule in simulated annealing. For any $0 < \alpha < 1$ and $\gamma > 0$, we have

$$A_i^t \to 0, r_i^t \to r_i^0, as\, t \to \infty. \tag{5}$$

In the simplest case, we can use $\alpha = \gamma$, and we have used $\alpha = \gamma = 0.9$ to 0.98 in our simulations [6].

3 Genetic Algorithms

Genetic algorithms (GAs) emulate genetic evolution. The characteristics of individuals are therefore expressed using genotypes. The original form of the GA, as illustrated by John Holland in 1975, had distinct features: (1) a bit string representation, (2) proportional selection, and (3) cross-over as the primary method to produce new individuals. Since then, several variations to the original Holland GA have been developed, using different representation schemes, selection, cross-over, mutation and elitism operators [3].

3.1 Representation

The classical representation scheme for GAs is a binary vector of fixed length. In the case of an n_x-dimensional search space, each individual consists on n_x variables with each variable encoded as a bit string. If variables have binary values, the length of each chromosome is n_x bits. In the case of nominal-valued variables, each nominal value can be encoded as an n_d-dimensional bit vectors where 2nd is the total numbers of discrete nominal values for that variable. Each n_d-bit string represents a different nominal value. In the case of continuous-valued variables, each variable should be mapped to an n_d-dimensional bit vector,

$$\phi : R \to (0, 1)^{n_d} \tag{6}$$

The range of continuous space needs to be restricted to a finite range, $[x_{min}, x_{max}]$. Using standard binary decoding, each continuous variable x_{ij} of chromosome x_i is encoded using a fixed length bit string.

GAs have also been developed that use integer or real-valued representations and order-based representations where the order of variables in a chromosome plays an important role. Also, it is not necessary that chromosomes be of fixed length [3].

3.2 Crossover Operations

Several crossover operators have been developed for GAs depending on the format in which individuals are represented. For binary representations, uniform crossover, one-point crossover and two-point crossover are the most popular:

- **Uniform Crossover**, where corresponding bit positions are randomly exchanged between the two parents to produce two offspring.
- **One-Point Crossover**, where a random bit position is selected, and the bit substrings after the selected bit are swapped between the two parents to produce two offspring.
- **Two-Point Crossover**, where two bit positions are randomly selected and the bit substrings between the selected bit positions are swapped between the two parents.

For continuous valued genes, arithmetic crossover can be used:

$$x_{ij} = r_j x_{1j} + (1.0 - r_j) x_{2j} \tag{7}$$

where $r_j \sim U(0, 1)$ and x_i is the offspring produced from parents x_1 and x_2 [3].

3.3 Mutation

The mutation scheme used in a GA depends on the representation scheme. In the case of bit string representations,

Random Mutation, randomly negates bits, while
In-Order Mutation, performs random mutation between two randomly selected bit positions.

For discrete genes with more than two possible values that a gene can assume, random mutation selects a random value from the finite domain of the gene. In the case of continuous valued genes, a random value sampled from a Gaussian distribution with zero mean and small deviation is usually added to the current gene value. As an alternative, random noise can be sampled from a Cauchy distribution [3].

4 Benchmark Mathematical Functions

This section lists a number of the benchmark mathematical functions used to evaluate the performance of the optimization algorithms.

In the area of optimization mathematical functions have been used to test different methods: Parallel Particle Swarm Optimization with Parameters Adaptation Using Fuzzy Logic [20], An improved evolutionary method with fuzzy logic for combining Particle Swarm Optimization and Genetic Algorithms [12], Optimal design of fuzzy classification systems using PSO with dynamic parameter adaptation through fuzzy logic [19], Parallel Particle Swarm Optimization with Parameters Adaptation Using Fuzzy Logic [16], and others in the literature.

The mathematical functions are defined below:

- **Sphere**

$$f(x) = \sum_{j=1}^{n_x} x_j^2 \qquad (8)$$

Witch $x_j \in [-100, 100]$ and $f^*(x) = 0.0$

- **Rosenbrock**

$$f(x) = \sum_{j=1}^{n_z/2} [100(x_{2j} - x_{2j-1}^2)^2 + (1 - x_{2j-1})^2] \qquad (9)$$

Witch $x_j \in [-2.048, 2.048]$ and $f^*(x) = 0.0$

- **Rastrigin**

$$f(x) = \sum_{j=1}^{n_x} (x_j^2 - 10\cos(2\pi x_j) + 10) \qquad (10)$$

With $x_j \in [-5.12, 5.12]$ and $f^*(x) = 0.0$

- **Ackley**

$$f(x) = -20e^{-0.2\sqrt{\frac{1}{n_x}\sum_{j=1}^{n_x} x_j^2 - \frac{1}{e^{n_x}}\sum_{j=1}^{n_x}\cos(2\pi x_j)}} + 20 + e \qquad (11)$$

With $x_j \in [-30, 30]$ and $f^*(x) = 0.0$

- **Zakharov**

$$f(x) = \sum_{i=1}^{n} x_i^2 + (\sum_{i=1}^{n} 0.5ix_i)^2 + (\sum_{i=1}^{n} 0.5ix_i)^4 \qquad (12)$$

Witch $x_i \in [-5, 10]$ and $f^*(x) = 0.0$

- **Sum Square**

$$f(x) = \sum_{i=1}^{n} ix_i^2 \tag{13}$$

Witch $x_i \in [-2, 2]$ and $f^*(x) = 0.0$

The mathematical functions were integrated directly into the code bat algorithm and genetic algorithm.

5 Results Between GA and the Bat Algorithm

In this section the Bat algorithm is compared against the genetic algorithm. In each of the algorithms, 6 Benchmark math functions were used separately for a dimension of 10 variables, and 30 tests were made for each function with different parameters in the algorithms.

The parameters in the Bat algorithm are as follows:

- Population size: 2–40 Bats
- Volume: 0.5–1
- Pulse frequency: 0.5–1
- Frequency min.: 0–2
- Frequency max.: 0–2

The parameters for the genetic algorithm are shown below:

- Number of Individuals: 4–40
- Selection: Stochastic, Remainder, Uniform, Roulette
- Crossover: Scattered, Single Point, Two Point, Heuristic, Arithmetic
- Mutation: Gaussian, Uniform

The results of the tests made with the De Jong's function for the two algorithms are shown in Table 1.

The results of the tests made with the Rosenbrock function for the two algorithms are shown in Table 2.

The results of the tests made with the Rastrigin function for the two algorithms are shown in Table 3.

The results of the tests made with the Ackley function for the two algorithms are shown in Table 4.

The results of the tests made with the Zakharov function for the two algorithms are shown in Table 5.

The results of the tests made with the Sum of Squares function for the two algorithms are shown in Table 6.

Table 1 Simulation results for the Jong's function

Bat algorithm		Genetic algorithm	
Numbers bats	Best	Population	Best
2	0.00857	4	0.00065
5	0.00360	5	0.01615
10	0.00002	10	0.02947
20	0.00013	20	0.12585
30	0.00001	30	0.05043
40	0.00000	40	0.00494

Table 2 Simulation results for the Rosenbrock function

Bat algorithm		Genetic algorithm	
Numbers bats	Best	Population	Best
2	0.00548	4	0.08984
5	0.27251	5	0.04515
10	0.30381	10	0.02647
20	0.13730	20	0.00875
30	0.38909	30	0.00122
40	0.25128	40	0.00024

Table 3 Simulation results for the Rastrigin function

Bat algorithm		Genetic algorithm	
Numbers bats	Best	Population	Best
2	0.07301	4	0.01493
5	0.06367	5	0.02538
10	0.04916	10	0.00098
20	0.00871	20	0.00586
30	0.00351	30	0.00146
40	0.04394	40	0.00171

Table 4 Simulation results for the Ackley function

Bat algorithm		Genetic algorithm	
Numbers bats	Best	Population	Best
2	0.00060	4	0.00376
5	0.00040	5	0.03481
10	0.00001	10	0.00068
20	0.00018	20	0.00000
30	0.00017	30	0.00249
40	0.00003	40	0.00110

Table 5 Simulation results for the Zakharov function

Bat algorithm		Genetic algorithm	
Numbers bats	Best	Population	Best
2	0.02818	4	0.00291
5	0.02152	5	0.00392
10	0.00010	10	0.00084
20	0.00005	20	0.01018
30	0.00000	30	0.00259
40	0.00000	40	0.00638

Table 6 Simulation results for the sum square function

Bat algorithm		Genetic algorithm	
Numbers bats	Best	Population	Best
2	0.01027	4	0.02947
5	0.00129	5	0.00958
10	0.00010	10	0.00084
20	0.00002	20	0.01018
30	0.00000	30	0.00259
40	0.00003	40	0.00638

In the comparative study of genetic algorithms and the effectiveness of the bat algorithm, the bat algorithm results with modification of parameters by trial and error are good but the rate of convergence of the genetic algorithm is much faster, the comparison was made with the original versions of the two algorithms with the recommended literature parameters.

6 Proposed Method

The Bat Algorithm has the characteristic of being one of the best to face problems of nonlinear global optimization. In this paper the enhancement of the bat algorithm using a fuzzy system is presented with the aim of dynamically setting some of the parameters in the algorithm. The goal is improving the performance of the algorithm against other metaheuristics in optimization problems by testing through the use of benchmark mathematical functions.

In the area of fuzzy logic for adapting parameters in metaheuristics we can find a similar work: Dynamic Fuzzy Logic Parameter Tuning for ACO and Its Application in TSP Problems [16].

Usually in the bat algorithm, the modification of the parameters is done by trial and error, modifying the parameters, which are wavelength λ, loudness (volume) A0, low frequency and high frequency. In the present work an implementation of a fuzzy system, which is responsible for setting any of these parameters dynamically in order to improve the performance of the algorithm achieving greater effectiveness is presented.

Once the modification in the bat algorithm is performed, tests were performed with benchmark mathematical functions to analyze its effectiveness. At the end, the original method is compared with the proposed method and also compared with the genetic algorithm, providing a more complete analysis of the effectiveness of the bat algorithm. Simulation results with the fuzzy bat algorithm outperform the traditional bat algorithm and genetic algorithms.

The general approach of the proposed bat algorithm method can be seen in Fig. 2.

The fuzzy system proposed is of Mamdani type because it is more common in this type of fuzzy control and the defuzzification method was the centroid. The membership functions are of triangular form in the inputs and outputs.

Also, the membership functions were chosen of triangular form based on past experiences in this type of fuzzy control. The fuzzy system consists of 9 rules.

In this section the comparison of the Bat algorithm is made against the fuzzy Bat Algorithm for each of the algorithms we consider 6 Benchmark math functions separately for a dimension of 10 variables, where 30 tests were performed for each function varying the parameters of the algorithms.

The results of the tests of the De Jong's function between the original method and the proposed one taking the best result of 30 experiments for each method are shown in Table 7.

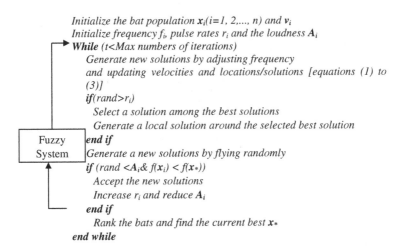

Fig. 2 Proposed scheme of the Bat Algorithm

Table 7 Simulation results for the De Jong's function

Bat algorithm		Fuzzy Bat algorithm	
Number of bats	Best	Numbers of bats	Best
40	0.000009	10	0.000001

The results of the tests of the Rosenbrock function between the original method and the proposed one taking the best result of 30 experiments for each method are shown in Table 8.

The results of the tests of the Rastrigin function between the original method and the proposed one taking the best result of 30 experiments for each method are shown in Table 9.

The results of the tests of the Acley function between the original method and the proposed one taking the best result of 30 experiments for each method are shown in Table 10.

The results of the tests of the function between the original method and the proposed one taking the best result of 30 experiments for each method are shown in Table 11.

Table 8 Simulation results for the Rosenbrock function

Bat algorithm		Fuzzy Bat algorithm	
Number of bats	Best	Number of bats	Best
2	0.005489	2	0.01275594

Table 9 Simulation results for the Rastrigin function

Bat algorithm		Fuzzy Bat algorithm	
Number of bats	Best	Number of bats	Best
30	0.003517	30	0.02243067

Table 10 Simulation results for the Ackley function

Bat algorithm		Fuzzy Bat algorithm	
Number of bats	Best	Number of bats	Best
10	0.000014	30	0

Table 11 Simulation results for the Zakharov function

Bat algorithm		Fuzzy Bat algorithm	
Number of bats	Best	Number of bats	Best
30	0.000005	30	0

Table 12 Simulation results for the sum squared function

Bat algorithm		Fuzzy Bat algorithm	
Number of bats	Best	Number of bats	Best
30	0.000006	10	0.00899885

The results of the tests of the Sum Squared function between the original method and the proposed one taking the best result of 30 experiments for each method are shown in Table 12.

7 Bat Algorithm Apply to the Inverted Pendulum

In this section we consider the application of the bat algorithm to the design of a controller to demonstrate the effectiveness of the algorithm. In the literature we can find that the following works have applied this method: Optimal Power Dispatch Using Bat Algorithm [2], Application of an improved SVR based Bat algorithm for short-term price forecasting in the Iranian Pay-as-Bid electricity market [18], Bat algorithm for topology optimization in microelectronic applications [24], Optimal placement and sizing of DER's with load models using BAT algorithm [21], Optimal Partial-Retuning of Decentralised PI Controller of Coal Gasifier Using Bat Algorithm [11], Design of Optimal Membership Functions for Fuzzy Controllers of the Water Tank and Inverted Pendulum with PSO Variants [4], and others in the literature.

In this section we present the application of the bat algorithm in the development of the fuzzy controller of the inverted pendulum. The proposed methodology is to perform the analysis of the problem of the inverted pendulum, which consists of the system in this example consists of an inverted pendulum mounted to a motorized cart. The inverted pendulum system is an example commonly found in control system textbooks and research literature. Its popularity derives in part from the fact that it is unstable without control, that is, the pendulum will simply fall over if the cart isn't moved to balance it. Additionally, the dynamics of the system are non-linear. The objective of the control system is to balance the inverted pendulum by applying a force to the cart that the pendulum is attached to. A real-world example that relates directly to this inverted pendulum system is the attitude control of a booster rocket at takeoff.

The Bat algorithm would take care of finding the optimal parameter values for some of the inverted pendulum problem, the end of the integration results with the original parameters recommended in the literature and the results obtained with the application of the algorithm will be shown.

8 Conclusions

In the simulation analysis for the comparative study of genetic algorithms and the effectiveness of the bat algorithm, the rate of convergence of the genetic algorithm is faster, the comparison was made with the original versions of the two algorithms with the recommended literature parameters, this conclusion is based on 6 Benchmark math functions results may vary according to mathematical or depending on the values set in the parameters of the algorithm.

In Sect. 6 the analysis for the comparative study of Bat Algorithm and proposed modification of Bat Algorithm, we find that there are promising results, but we read to continue improving the algorithm.

The application of the bat algorithm to various problems has a very wide field where the revised items its effectiveness is demonstrated in various applications, their use can be mentioned in the processing digital pictures, search for optimal values, neural networks, and many applications.

Acknowledgments We would like to express our gratitude to the CONACYT and Tijuana Institute of Technology for the facilities and resources granted for the development of this research.

References

1. Alemu, T., Mohd, F.: Use of Fuzzy Systems and Bat Algorithm for Exergy Modeling in a Gas Turbine Generator. TamiruAlemu Lemma, Department of Mechanical Engineering, Malaysia, (2011)
2. Biswal, S., Barisal, A.K., Behera, A., Prakash, T.: Optimal Power Dispatch Using Bat Algorithm, Department of Electrical Engineering., VSSUT, Burla, India, (2013)
3. Engelbrecht, A.: Fundamentals of Computational Swarm Intelligence. A. P. Engelbrecht University of Pretoria South Africa, Wiley. pp. 25–26 and 66–70, (2005)
4. Fierro, R., Castillo, O., Valdez, F., Cervantes, L.: Design of optimal membership functions for fuzzy controllers of the water tank and inverted pendulum with PSO variants. In: IFSA World Congress and NAFIPS Annual Meeting (IFSA/NAFIPS), 2013 Joint
5. Gandomi, A., Yang, X.: Chaotic Bat Algorithm. The University of Akron, Department of Civil Engineering, USA (2013)
6. Goel, N., Gupta, D., Goel, S.: Performance of Firefly and Bat Algorithm for Unconstrained Optimization Problems, Department of Computer Science Maharaja Surajmal. Institute of Technology GGSIP university C-4 Janakpuri, New Delhi, India (2013)
7. Hasançebi, O., Carbas, S.: Bat inspired Algorithm For Discrete Size Optimization Of Steel Frames. Department of Civil Engineering, Middle East Technical University, 06800 Ankara, Turkey, (2013)
8. Hasançebi, O., Teke, T., Pekcan, O.: A Bat-Inspired Algorithm For Structural Optimization. Middle East Technical University Department of Civil Engineering, Ankara, Turkey (2013)
9. Kashi, S., Minuchehr, A., Poursalehi, N., Zolfaghari, A.: Bat Algorithm For The Fuel Arrangement Optimization of Reactor Core. ShahidBeheshti University, Nuclear Engineering Department, Tehran, Iran (2013)

10. Khan, K., Sahai, A. A.: Comparison of BA, GA, PSO, BP and LM for Training Feed forward Neural Networks in e-Learning Context. Department of Computing and Information Technology, University of the West Indies, St. Augustine, Trinidad And Tobago, (2012)
11. Kotteeswaran, R., Sivakumar, L.: Optimal Partial-Retuning of Decentralised PI Controller of Coal Gasifier Using Bat Algorithm. Swarm, Evolutionary, and Memetic Computing, Springer, pp. 750–761, (2013)
12. Melin, P., Olivas, F., Castillo, O., Valdez, F., Soria, J., Garcia, J.: Optimal design of fuzzy classification systems using PSO with dynamic parameter adaptation through fuzzy logic. Expert Syst. Appl. **40**(8), 3196–3206 (2013)
13. Mishra, S., Shaw, K., Mishra, D.: A New Meta-heuristic Bat Inspired Classification Approach for Microarray Data. Siksha O Anusandhan Deemed to be University, Institute of Technical Education and Research, Bhubaneswar, Odisha, India (2011)
14. Musikapun, P., Pongcharoen, P.: Solving Multi-Stage MultiMachine Multi-Product Scheduling Problem Using Bat Algorithm. Faculty of Engineering, Naresuan University, Department of Industrial Engineering, Thailand (2012)
15. Nakamura, R., Pereira, L., Costa, K., Rodrigues, D., Papa, J.: BBA: A Binary Bat Algorithm for Feature Selection. Department of Computing Sao Paulo State University Bauru, Brazil, (2012)
16. Neyoy, H., Castillo, O., Soria, J.: Dynamic fuzzy logic parameter tuning for ACO and its application in TSP problems. In: Recent Advances on Hybrid Intelligent Systems, pp. 259–271, (2013)
17. Rodrigues, D., Pereira, L., Nakamura, R., Costa, K., Yang, X., Souza, A., Papa, J. P.: A Wrapper Approach for Feature Selection Based on Bat Algorithm and Optimum-Path Forest. Department of Computing, Universidade Estadual Paulista, Bauru, Brazil, (2013)
18. Taherian, H., NazerKakhki, I., Aghaebrahimi, M.: Application of an Improved SVR Based Bat Algorithm for Short-Term Price Forecasting in the Iranian Pay-as-Bid Electricity Market. University of Birjand, Birjand, Department of Electrical and Computer Engineering, Iran (2013)
19. Valdez, F., Melin, P., Castillo, O.: Evolutionary method combining particle swarm optimization and genetic algorithms using fuzzy logic for decision making. In: Proceedings of the IEEE International Conference on Fuzzy Systems, pp. 2114–2119, (2009)
20. Valdez, F., Melin, P., Castillo, O.: Parallel Particle Swarm Optimization witch Parameters Adaptation Using Fuzzy Logic. In: Batyrshin, I., González Mendoza, M. (eds.): MICAI 2012, Part II, LNAI 7630, pp. 374–385, 2012. Springer Berlin Heidelberg (2012)
21. Yammani, C., Maheswarapu, S., Sailaja Kumari, M.: Optimal Placement and Sizing of DER's with Load Models Using BAT Algorithm. Electrical Engineering Department, National Institute of Technology, Warangal, India (2013)
22. Yang, X.: A New Metaheuristic Bat-Inspired Algorithm. Department of Engineering, University of Cambridge, Trumpington Street, Cambridge CB2 1PZ, UK, (2010)
23. Yang, X.: Bat Algorithm: Literature Review and Applications. School of Science and Technology, Middlesex University, The Burroughs, London NW4 4BT, United Kingdom, (2013)
24. Yang, X., Karamanoglu, M., Fong, S.: Bat algorithm for topology optimization in microelectronic applications. School of Science and Technology, Middlesex University, Hendon Campus, London NW4 4BT, UK, (2012)
25. Yuanbin, M., Xinquan, Z., Sujian, X.: Local Memory Search Bat Algorithm for Grey Economic Dynamic System. Statistics and Mathematics Institute, (2013)

Optimization of Benchmark Mathematical Functions Using the Firefly Algorithm with Dynamic Parameters

Cinthya Solano-Aragón and Oscar Castillo

Abstract Nature-inspired algorithms are more relevant today, such as PSO and ACO, which have been used in several types of problems such as the optimization of neural networks, fuzzy systems, control, and others showing good results [1–5]. There are other methods that have been proposed more recently, the firefly algorithm is one of them, this paper will explain the algorithm and describe how it behaves. In this paper the firefly algorithm was applied in optimizing benchmark functions and comparing the results of the same functions with genetic algorithms.

Keywords Genetic algorithms · Firefly algorithm · Benchmark functions · Optimization

1 Introduction

Optimization is the process of adjusting the inputs to or characteristics of a device, mathematical process, or experiment to find the minimum or maximum output or result (Fig. 1). The input consists of variables; the process or function is known as the cost function, objective function, or fitness function; and the output is the cost or fitness. If the process is an experiment, then the variables are physical inputs to the experiment [6].

The main problem of optimization is to find the values of the variables of a function to be optimized. These types of problems exist in many disciplines. Despite the fact that there are many methods of solution, there are many problems that need special attention and are difficult to solve using deterministic solution methods. In contrast to the deterministic algorithms, meta-heuristic methods are not affected by the behavior of the optimization problem. This makes the algorithms more widely usable.

C. Solano-Aragón · O. Castillo (✉)
Tijuana Institute of Technology, Tijuana, Mexico
e-mail: ocastillo@tectijuana.mx

© Springer International Publishing Switzerland 2015
O. Castillo and P. Melin (eds.), *Fuzzy Logic Augmentation of Nature-Inspired Optimization Metaheuristics*, Studies in Computational Intelligence 574,
DOI 10.1007/978-3-319-10960-2_5

81

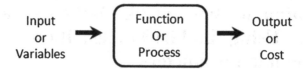

Fig. 1 Diagram of a function or process that is to be optimized. Optimization varies the input to achieve a desired output

The rest of the paper is organized as follows. Section 2 we describe the Firefly Algorithm developed by Xin-She Yang. In Sect. 3, talks about methodology for parameter adaptation. In Sect. 4, shows the experimentation with benchmark mathematical functions. Finally Sect. 5, shows the conclusions.

2 Firefly Algorithm

The Firefly Algorithm (FA) is a meta-heuristic, nature-inspired, optimization algorithm which is based on the social (flashing) behavior of fireflies. The flashing light of fireflies is an amazing sight in the summer sky in the tropical and temperate regions. The primary purpose for a firefly's flash is to act as a signal system to attract other fireflies. In addition, flashing may also serve as a protective warning mechanism [7].

For simplicity, the flashing characteristics of fireflies are idealized in the following three rules [8–10]:

- All fireflies are unisex, so that one firefly is attracted to other fireflies regardless of their sex.
- Attractiveness is proportional to their brightness, thus for any two flashing fireflies, the less bright one will move towards the brighter one. The attractiveness is proportional to the brightness and they both decrease as their distance increases. If no one is brighter than a particular firefly, it moves randomly.
- The brightness of a firefly is affected or determined by the landscape of the objective function to be optimized.

2.1 Attractiveness

The form of the attractiveness function of a firefly is the following monotonically decreasing function [8]:

$$\beta(r) = \beta_0 e^{-\gamma r^m} \quad (m \geq 1) \tag{1}$$

where r is the distance between any two fireflies, β_0 is the attractiveness at $r = 0$ and γ is a fixed light absorption coefficient.

2.2 Distance

The distance between any two fireflies i and j at X_i and X_j, respectively, is the Cartesian distance as follows:

$$r_{ij} = \| X_i - X_j \| = \sqrt{\sum_{k=1}^{d} \left(x_{i,k} - x_{j,k} \right)^2} \qquad (2)$$

where $x_{i,k}$ is the kth component of the spatial coordinate X_i of ith firefly and d is the number of dimensions.

2.3 Movement

The movement of a firefly i is attracted to another more attractive (brighter) firefly j is determined by following equation:

$$x_i^{t+1} = x_i^t + \beta_0 e^{-\gamma r_{ij}^2} \left(x_j^t - x_i^t \right) + \alpha_t \in_i^t \qquad (3)$$

where the second term is due to the attraction while the third term is randomization with α being the randomization parameter. *rand* is a random number generator uniformly distributed in [0, 1]. For most cases in the implementation, $\beta_0 = 1$ *and* $\alpha \in [0, 1]$.

3 Methodology for Parameter Adaptation

Because in the Firefly Algorithm, there are two important issues: the variation of light intensity and formulation of the attractiveness, we decided to use the parameters that affect to implement changes in its evolution expecting better results.

Here we use fuzzy logic and describe the system used.

In Fig. 2 we can see the parameters can will be a change in the evolution of algorithm before showing the different variations of FIS that we develop in search of better results.

The design of the input variables can be appreciated in Figs. 3, 4 and 5, which show the inputs iteration, diversity, and error respectively, each input is granulated into three triangular membership functions.

For the output variables, the recommended values for β are between 0 and 1 and γ are between 0.1 and 10, so that the output variables were designed using this range of values. Each output is granulated in five triangular membership functions,

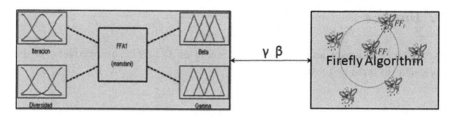

Fig. 2 Scheme of dynamic parameters

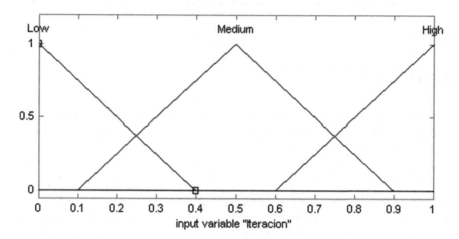

Fig. 3 Input 1: iteration

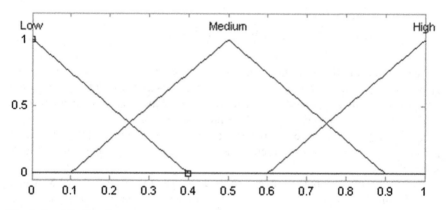

Fig. 4 Input 2: diversity

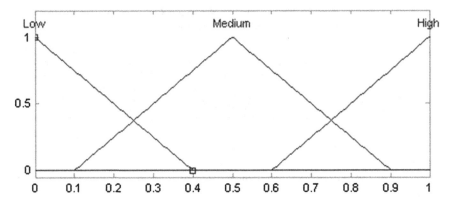

Fig. 5 Input 3: error

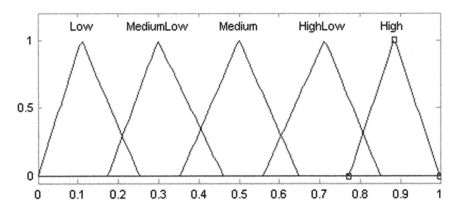

Fig. 6 Output 1: β (Beta)

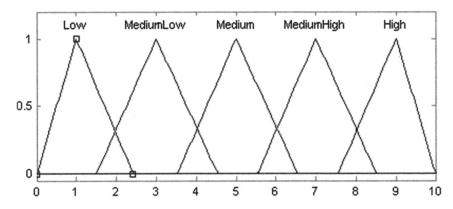

Fig. 7 Output 2: γ (Gamma)

1. If (Iteracion is Low) and (Diversidad is Low) then (Beta is Medium)(Gamma is Medium) (1)
2. If (Iteracion is Low) and (Diversidad is Medium) then (Beta is MediumLow)(Gamma is MediumHigh) (1)
3. If (Iteracion is Low) and (Diversidad is High) then (Beta is MediumHigh)(Gamma is MediumLow) (1)
4. If (Iteracion is Medium) and (Diversidad is Low) then (Beta is MediumLow)(Gamma is MediumLow) (1)
5. If (Iteracion is Medium) and (Diversidad is Medium) then (Beta is Medium)(Gamma is MediumLow) (1)
6. If (Iteracion is Medium) and (Diversidad is High) then (Beta is MediumHigh)(Gamma is MediumLow) (1)
7. If (Iteracion is High) and (Diversidad is Low) then (Beta is Low)(Gamma is Low) (1)
8. If (Iteracion is High) and (Diversidad is Medium) then (Beta is MediumHigh)(Gamma is Low) (1)
9. If (Iteracion is High) and (Diversidad is High) then (Beta is High)(Gamma is Low) (1)

Fig. 8 Rules of fuzzy system

the design of the output variables can be seen in Figs. 6 and 7, β (Beta) and γ (Gamma) respectively.

Having defined the possible input variables, it was decided to combine them to generate different fuzzy systems for dynamic adjustment of β and γ (Fig. 8).

4 Experimentation with Benchmark Mathematical Functions

In the field of evolutionary computation, it is common to compare different algorithms using a large test set, especially when the test set involves function optimization. However, the effectiveness of an algorithm against another algorithm cannot be measured by the number of problems that it solves better. If we compare two searching algorithms with all possible functions, the performance of any two algorithms will be, on average, the same. As a result, attempting to design a perfect test set where all the functions are present in order to determine whether an algorithm is better than other for every function. This is the reason why, when an algorithm is evaluated, we must look for the kind of problems where its performance is good, in order to characterize the type of problems for which the algorithm is suitable. In this way, we have made a previous study of the functions to be optimized for constructing a test set with six benchmark functions and a better selection. This allows us to obtain conclusions of the performance of the algorithm depending on the type of function. The mathematical functions analyzed in this paper are in the table. The functions of Table 1 were evaluated considering 20 variables [11–14] .

We performed additional experiments using the functions contained in Haupt and Haupt [15], we present the parameters used in Table 2 for all experiments. In Table 3 we can see the results of first experiments using Firefly Algorithm (Simple FA) and Fuzzy Firefly Algorithm (Fuzzy FA). In Table 4 we find the results of the functions contained in the book of Haupt and Haupt [15] using Firefly Algorithm (Simple FA) and Fuzzy Firefly Algorithm (Fuzzy FA).

Table 1 Mathematical functions

Function	Expression
Rastrigin	$f(x) = 10n + \sum\limits_{i=1}^{n}(x^2 - 10\cos(2\pi x_i))$
Rosenbrock	$f(x) = \sum\limits_{i=1}^{n-1}\left[100\left(x_i^2 - x_i^2 + 1\right)^2 + (x_i - 1)^2\right]$
Ackley	$f(x) = 20 + e - 20e^{-1/5}\sqrt{1/n\sum\limits_{i=1}^{n}x_i^2} - e^{1/n}\sum\limits_{i=1}^{n}\cos(2\pi x_1)$
Shubert	$f(x) = \left(\sum\limits_{i=1}^{5}i\cos((i+1)x_1) + i\right)\left(\sum\limits_{i=1}^{5}i\cos((i+1)x_2 + i\right)$
Sphere	$f(x) = \sum\limits_{i=1}^{n}x_i^2$
Griewank	$f(x) = \sum\limits_{i=1}^{n}\frac{x_i^2}{4000} - \prod\limits_{i=1}^{n}\cos\left(x_i/\sqrt{i}\right) + 1$

Table 2 Parameters for each method

Parameter	Simple FA	Fuzzy FA
Population	25	25
Iterations	20	20
β (Beta)	0.20	Dynamic
γ (Gamma)	1.0	Dynamic

Table 3 Simulation results

Mathematical functions	Minimum	Simple FA	Fuzzy FA
Rastrigin	0	3.47E−02	7.28E−03
Rosenbrock	0	7.39E−03	7.50E−03
Ackley	0	1.89E−02	1.19E−02
Shubert	−186.73	−186.584	−186.597
Sphere	0	7.55E−04	6.45E−03
Griewank	0	1.67E−02	5.58E−03

Table 4 Simulation results of Haupt, R., and Haupt, S

Function	Minimum	Simple FA	Fuzzy FA
f1	1	0.9598	0.9905
f2	0	1.64E−02	6.78E−03
f3	1	0.9875	0.9874
f4	0	5.22E−03	1.80E−02
f5	0–1	3.46E−02	2.08E−02
f6	−100.22	−99.9847	−100.1064
f7	−18.5547	−18.2410	−18.51348
f8	−18.5547	−17.9456	−18.5374
f9	0	6.70E−02	3.18E−04
f10	0	1.63E−02	3.34E−03
f11	0	5.69E−02	2.74E−02
f12	−0.5231	−0.5121	−0.5218
f13	0	6.83E−03	6.83E−03
f14	−0.3356	−0.3312	−0.3324
f15	−16.947	−15.9856	−16.5745
f16	−23.806	−23.5312	−23.7412

5 Conclusion

The behavior of the algorithm along the conducting experiments was stable and do not show a large expected compared with other algorithms that have been used for solving these problems despite the use of dynamic parameters.

In the previous work [16] that had been done we used the algorithm without any modification Firefly no improvement and was likewise very near but still searched results, we think you could still improve; so we decided to make the change parameters dynamically using fuzzy logic.

Find better solutions in most of the experiments, we used a total of 22 math functions and experiments performed 10 each; we can conclude that the firefly algorithm is a good method for solving functions.

References

1. Goldberg, D.E.: Genetic Algorithms in Search, Optimization and Machine Learning, Reading, Mass, Addison Wesley, Reading (1989)
2. Melendez, A., Castillo, O.: Evolutionary optimization of the fuzzy integrator in a navigation system for a mobile robot. Recent Adv. Hybrid Intell. Syst. 21–31 (2013)
3. Rodriguez Vázquez, K.: Multiobjective Evolutionary Algorithms in Non-linear System Identification. Automatic Control and Systems Engineering, The University of Sheffield, Sheffield, p. 185 (1999)

4. Astudillo, L., Melin, P., Castillo, O.: Optimization of a fuzzy tracking controller for an autonomous mobile robot under perturbed torques by means of a chemical optimization paradigm. Recent Adv. Hybrid Intell. Syst. 3–20 (2013)
5. Cervantes, L., Castillo, O.: Genetic optimization of membership functions in modular fuzzy controllers for complex problems. Recent Adv Hybrid Intell. Syst. 51–62 (2013)
6. Holland, H.: Adaptation in Natural and Artificial Systems. University of Michigan Press, Ann Arbor (1975)
7. Yang, X.S.: Nature-Inspired Metaheuristic Algorithms. Luniver Press, Europe (2008)
8. Yang, X.S.: Firefly algorithms for multimodal optimization. In: Stochastic Algorithms Foundations and Applications (SAGA'09). Lecture Notes in Computing Sciences, Vol. 5792. , Springer, New York, pp. 169–178 (2009)
9. Yang, X.S.: Firefly algorithm stochastic test functions and design optimization. Int. J. Bio-Inspired Comput. 2(2), 78–84 (2010)
10. Yang, X.S.: Firefly algorithm, lévy flights and global optimization. In: Bramer, M., Ellis, R., Petridis, M. (eds.) Research and Development in Intelligent Systems, Vol. XXVI, pp. 209–218. Springer, London (2010)
11. Valdez, F., Melin, P.: Comparative study of particle swarm optimization and genetic algorithms for mathematical complex functions. J. Autom. Mob. Robot. Intell. Syst. (JAMRIS) 2, 43–51 (2008)
12. Valdez, F., Melin, P., Castillo, O.: An improved evolutionary method with fuzzy logic for combining particle swarm optimization and genetic algorithms. Appl. Soft Comput. 11(2), 2625–2632 (2011)
13. Valdez, F., Melin, P., Castillo, O.: Bio-inspired optimization methods on graphic processing unit for minimization of complex mathematical functions. Recent Adv. Hybrid Intell. Syst. 313–322 (2013)
14. Melin, P., Olivas, F., Castillo, O., Valdez, F., Soria, J., García, J.M.: Valdez: optimal design of fuzzy classification systems using PSO with dynamic parameter adaptation through fuzzy logic. Expert Syst. Appl. 40(8), 3196–3206 (2013)
15. Haupt, R., Haupt, S.: Practical genetic algorithms 2nd ed. A Wiley-Interscience Publication (1998)
16. Solano-Aragon, C., Castillo, O.: Optimization of benchmark mathematical functions using the firefly algorithm. Recent Adv. Hybrid Approaches Designing Intell. Syst. 177–189 (2013)

Cuckoo Search via Lévy Flights and a Comparison with Genetic Algorithms

Maribel Guerrero, Oscar Castillo and Mario García

Abstract The purpose of this paper is to present a brief literature review of the cuckoo search algorithm (CS) and analyze its behavior by applying it to a set of benchmark mathematical functions. CS is a stochastic algorithm, inspired by the nature of a family bird called Cuckoo. CS algorithms are reinforced with Lévy flights to analyze the search space in a successful manner. We performed a comparison of Cuckoo Search (CS) and Genetic Algorithm (GA), these algorithms were tested on five mathematical functions for analysis.

Keywords Cuckoo search algorithm · Genetic algorithm · Levy flights

1 Introduction

In this paper we describe the Cuckoo Search Algorithm via Lévy flights with the intention of making a comparison with Genetic Algorithms. Based on the performed simulations, in Sect. 5 we present tables with the optimization results obtained after applying the CS and GA to mathematical functions.

The comparative study was performed in order to observe the behavior presented by the CS algorithm with 5 mathematical functions and with different number of variables.

Cuckoo Search (CS) is one of the latest nature-inspired meta-heuristic algorithms, developed in 2009 by Xin-She Yang of Cambridge University and Suash Deb of C. V. Raman College of Engineering. CS is based on the brood parasitism of some cuckoo species. In addition, this algorithm is enhanced by the so-called Lévy flights, rather than by simple isotropic random walks.

M. Guerrero · O. Castillo (✉) · M. García
Tijuana Institute of Technology, Tijuana, B.C., Mexico
e-mail: ocastillo@tectijuana.mx

© Springer International Publishing Switzerland 2015
O. Castillo and P. Melin (eds.), *Fuzzy Logic Augmentation of Nature-Inspired Optimization Metaheuristics*, Studies in Computational Intelligence 574,
DOI 10.1007/978-3-319-10960-2_6

91

Cuckoo Search has been applied in many areas of computational intelligence and optimization. For example, in engineering design applications, cuckoo search has superior performance over other algorithms for a range of continuous optimization problems such as the spring design and welded beam design problems [7, 8, 21].

Other works include a discrete cuckoo search algorithm proposed by Tein and Ramli [15] to solve nurse scheduling problems. Another one is the modified the cuckoo search via Lévy flights proposed by Yang and Deb [20] Lévy flights reinforce the CS algorithm, due to the behavior of some birds and fruit flies. Zheng and Zhou [23] provided a variant of Cuckoo Search using Gaussian process. Yang and Deb [22] proposed the MultiObjective Cuckoo Search (MOCS) for design engineering applications.

There are various applications including: Interesting results were obtained using Cuckoo Search for training neural networks, as shown Valian et al. [17]. The Cuckoo Search has also been used to generate independent paths for software testing and test data generation [3, 12, 14]. In the context of data fusion and wireless sensor network, Cuckoo Search has been shown to be very efficient [4, 5]. Furthermore, Vazquez [18] used Cuckoo Search to train spiking neural network models, while Chifu et al. [1] optimized semantic web service composition processes using Cuckoo Search.

The paper is organized as follows: in Sect. 2 the description of the Cuckoo Search Algorithm, its variants and the pseudo code are presented, in Sect. 3 a description of the Genetic Algorithm, in Sect. 4 the description of the mathematical functions is presented, in Sect. 5 we can find the results of the simulations, in Sect. 6 the conclusions from the analysis of results obtained after applying mathematical functions are presented.

2 Cuckoo Search Algorithm

The Cuckoo is a fascinating bird, not only because of the beautiful sound it can make, but also because of their aggressive reproduction strategy. Some species such as the Ani and Guira cuckoos lay their eggs in communal nests, though they may remove others' eggs to increase the hatching probability of their own eggs. Quite a number of species engage the obligate brood parasitism by laying their eggs in the nests of other host birds (often other species) [19].

For simplicity in describing the Cuckoo Search, we now use the following three idealized rules [13]:

1. Each cuckoo lays one egg at a time, and dumps its egg in randomly chosen nest.
2. The best nests with high quality of eggs will carry over to the next generations.

3. The number of available host nests is fixed, and the egg laid by a cuckoo is discovered by the host bird with a probability pa in [0, 1]. In this case, the host bird can either throw the egg away or abandon the nest, and build a completely new nest. For simplicity, this last assumption can be approximated by the fraction pa of the n nests are replaced by new nests (with new random solutions).

As a further approximation, this last assumption can be approximated by replacing a fraction pa of the n host nests with new nests (with new random solutions). For a maximization problem, the quality or fitness of a solution can simply be proportional to the value of the objective function. Other forms of fitness can be defined in a similar way to the fitness function in Genetic Algorithms.

The basic steps of the Cuckoo Search (CS) can be summarized as the pseudo code shown in Sect. 2.3.

2.1 Variants

The original Cuckoo Search was first tested using numerical function optimization benchmarks. Usually, this kind of problems represents a test bed for new developed algorithms. In line with this, standard benchmark function suites [9, 21] have been developed in order to make comparison between algorithms as fair as possible. For example, some original studies in this area are:

- Cuckoo Search via Lévy flights [20].
- An efficient Cuckoo Search algorithm for numerical function optimization [11].
- Multimodal function optimization [10].

Cuckoo search can deal with multimodal problems naturally and efficiently. However, researchers have also attempted to improve its efficiency further so as to obtained better solutions than those in the literature [6], and one such study that is worth mentioning is by Jamil and Zepernick [10].

2.2 Lévy Flights

On the other hand, various studies have shown that the flight behavior of many animals and insects has demonstrated the typical characteristics of Lévy flights. A recent study by Reynolds and Frye shows that fruit flies or Drosophila melanogaster, explore their landscape using a series of straight flight paths punctuated by a sudden 90° turn, leading to a Lévy-flight-style intermittent scale free search pattern.

Studies on human behavior such as the Ju/'hoansi hunter-gatherer foraging patterns also show the typical feature of Lévy flights. Even light can be related to Lévy flights. Subsequently, such behavior has been applied to optimization and optimal search, and preliminary results show its promising capability.

2.3 Pseudo Code for Cuckoo Search Algorithm

The pseudocode is as follows:

Begin
  Objective function f(x), x= $(x_1, ..., x_d)^T$
  Generate initial population of n host nests xi (i = 1, 2, ..., n)
while *(t <MaxGeneration) or (stop criterion)*
  Get a cuckoo randomly by Lévy flights evaluate its quality/fitness Fi
  Choose a nest among n (say, j) randomly
if*(Fi >Fj),*
replace j by the new solution;
end
  A fraction (P_a) of worse nests are abandoned and new ones are built;
  Keep the best solutions (or nests with quality solutions);
  Rank the solutions and find the current best
end while
Postprocess results and visualization
end

2.4 Generate a New Solution

When generating new solutions x(t + 1), for say a cuckoo *i*, a Lévy flight is performed using Eq. 1:

$$x_i^{(t+1)} = x_i + \alpha \otimes L\acute{e}vy(\lambda) \qquad (1)$$

where:
$x_i^{(t+1)}$ The new position,
x_i Current position
α Is the step size which should be related to the scales of the problem of interests, where $\alpha \geq 0$

The product $\otimes$ means entry-wise multiplications.

$Lévy(\lambda)$ The probability distribution
λ It is a constant $(1 \leq \lambda \leq 3)$.
t The number of current generation (Time)

In general, a random walk is a Markov chain whose next status/location only depends on the current location (the first term in the above Eq. 1) and the transition probability (the second term).

The Lévy flight essentially provides a random walk while the random step length is drawn from a Lévy distribution, in Eq. 2 we can see that distribution:

$$Lévy \sim u = t^{-\lambda}, \quad (1 < \lambda \leq 3), \tag{2}$$

Which has an infinite variance with an infinite mean. Here the steps essentially form a random walk process with a power law step-length distribution with a heavy tail. Some of the new solutions should be generated by Lévy walk around the best solution obtained so far, this will speed up the local search.

However, a substantial fraction of the new solutions should be generated by far field randomization and whose locations should be far enough from the current best solution, this will make sure the system will not be trapped in a local optimum.

3 Genetic Algorithms

The Genetic Algorithm (GA) is an optimization and search technique based on the principles of genetics and natural selection. A GA allows a population composed of many individuals to evolve under specified selection rules to a state that maximizes the "fitness" (i.e., minimizes the cost function).

Genetic Algorithm (GA), is introduced by John Holland from the University of Michigan initiated his work on genetic algorithms at the beginning of the 1960s, is the powerful stochastic algorithm based on the principles of natural selection and natural genetic, applied in optimization problems and machine learning.

In GA maintains a population of individuals, each individual is represented by a strings or chromosomes and y probabilistically modifies the population by some genetic operators such as selection, crossover and mutation, with the intent of seeking a near-optimal solution to the problem.

The main driving operators of a GA are selection (to model survival of the fittest) and recombination through application of a crossover operator (to model reproduction).

3.1 Representation—the Chromosome

An individual in a GA is usually represented by a vector of fixed length with values in each of their positions of 0 or 1. If we have an ND-dimensional search space, each individual consists of n variables with each variable encoded as a bit string; the length of each chromosome is ND bits. In the case of nominal-valued variables, each nominal value can be encoded as an ND-dimensional bit vector where 2ND is the total numbers of discrete nominal values for that variable. Each ND-bit string represents a different nominal value. In the case of continuous-valued variables, each variable should be mapped to an ND-dimensional bit vector.

$$\phi : R \to (0, 1)^{n_d} \tag{3}$$

The domain of continuous space needs to be restricted to a finite range, [xmin, xmax]. Using standard binary decoding, each continuous variable xij of chromosome xi is encoded using a fixed length bit string.

3.2 Crossover Operations

Several crossover operators have been developed to compute for GAs depending on the format in which individuals are represented:

- One-point crossover: Holland [2] suggested that segments of genes be swapped between the parents to create their offspring, and not single genes.
  A one-point crossover operator was developed that randomly selects a crossover point and the bit strings after that point are swapped between the two parents.
- Two-point crossover: In this case two bit positions are randomly selected, and the bit strings between these points are swapped.

Uniform Crossover, where corresponding bit positions are randomly exchanged between the two parents to produce two offspring

3.3 Mutation

For binary representations, the following mutation operators have been developed:

- Uniform (random) mutation, where bit positions are chosen randomly and the corresponding bit values negated.
- In-order mutation, where two mutation points are randomly selected and only the bits between these mutation points undergo random mutation.

- Gaussian mutation: the bit string that represents a decision variable be converted back to a floating-point value and mutated with Gaussian noise. For each chromosome random numbers are drawn from a Poisson distribution to determine the genes to be mutated.

4 Benchmark Mathematical Functions

To test the CS and GA algorithm, we use a set of 5 benchmark functions used to evaluate the performance of optimization algorithms were obtained [16], called F1 (Spherical Function), F2 (Rosenbrock Function), F3 (Ackley Function), F4 (Rastringin Function) and F5 (Griewank Function).

The functions were evaluated with 8, 16, 32, 64 and 128 dimensions.

Figure 1 the Eq. 4 represents the F1 function, the Eq. 5 represents the F2 function, so on.

The mathematical functions are shown below:

Spherical Function (F1)

$$f(x) = \sum_{j=1}^{n_x} x_j^2 \tag{4}$$

Witch $x_j \in [-5.12, 5.12]$ and $f^*(x) = 0.0$

Rosenbrock Function (F2)

$$f(x) = \sum_{j=1}^{n_z/2} [100(x_{2j} - x_{2j-1}^2)^2 + (1 - x_{2j-1})^2] \tag{5}$$

Witch $x_j \in [-5, 10]$ and $f^*(x) = 0.0$

Ackley Function (F3)

$$f(x) = 20 + e - 20e^{-1/5} \sqrt{\frac{1}{n} \sum_{j=1}^{n} x_j^2} - e^{\frac{1}{n} \sum_{j=1}^{n} \cos(2\pi x_j)} \tag{6}$$

With $x_j \in [-5, 30]$ and $f^*(x) = 0.0$

Rastringin Function (F4)

$$f(x) = 10n \sum_{j=1}^{n_x} [x_j^2 - 10\cos(2\pi x_j)] \tag{7}$$

With $x_j \in [-5.12, 5.12]$ and $f^*(x) = 0.0$

Griewank Function (F5)

Table 1 Parameters used in CS

Parameter	Value
Population size	100 nests
Pa	0.75 probability Pa discovered by the host bird
α	0.05 step size

Table 2 Parameters used in GA

Parameter	Value
Population size	100 individuals
Crossover (k1)	80 %
Mutation (k2)	5 %
Selection	Roulette

$$f(x) = 1 + \sum_{n=1}^{n_x} \frac{x_n^2}{40000} \prod_{n=1}^{N} \cos(x_n) \tag{8}$$

Witch $x_i \in [-600, 600]$ and $f^*(x) = 0.0$

5 Simulation Results

The test of the CS and GA was madean implementation in the Matlab programming language.

The implementation was developed for CS using a computer with processor Intel Core 2 Duo of 64 bits that works to a frequency of clock of 2.93 GHz, 4.00 GB of RAM Memory and Windows 7 Ultimate Operating System. Data for tests with GA were obtained from reference [16].

In the tables we can find the number of variables used (VARIABLE), the best result obtained (BEST), the average of 50 times (AVERAGE), the worst results obtained (WORST).

In Table 1,we can find the parameters used for all the tests performed for the CS algorithm.

In the Table 2, we can find the parameters used for all the tests performed for the GA algorithm.

5.1 Simulation Results with the Cuckoo Search Algorithm

In this Section we show tables from 7 to 12 where it can be appreciated that after executing the Cuckoo Search algorithm 50 times, with different number of

Table 3 Experimental results with CS for spherical Function (F1)

F1			
Variables	Best	Average	Worst
8	2.50E−25	4.74E−24	2.13E−23
16	2.25E−13	4.99E−13	1.05E−12
32	4.25E−07	8.92E−07	1.65E−06
64	0.001709	0.002822	0.00453256
128	0.227951	0.33637	0.43696553

Table 4 Experimental results with CS for Rosenbrock function (F2)

F2			
Variables	Best	Average	Worst
8	0.0017451	0.02512	0.27673769
16	0.1928658	3.752432	8.34213099
32	24.63969	28.39019	31.5022726
64	151.64964	230.5317	293.189601
128	2262.9706	2677.951	3431.05473

Table 5 Experimental results with CS for Ackley function (F3)

F3			
Variables	Best	Average	Worst
8	2.65E−07	1.65E−06	7.18E−06
16	0.00558276	0.01776329	0.04590806
32	0.14201793	0.49304933	1.02257225
64	2.66494517	3.07581938	3.726809
128	2.66494517	3.07581938	3.726809

Table 6 Experimental results with CS for Rastringin function (F4)

F4			
Variables	Best	Average	Worst
8	0.38091491	1.62804412	2.40289176
16	16.3558737	23.8306045	31.1494267
32	84.1365234	108.086428	127.969947
64	282.981597	346.106282	400.833459
128	812.083589	912.6544	993.524008

Table 7 Experimental results with CS for Griewank function (F5)

F5			
Variables	Best	Average	Worst
8	0.2534578	0.04387996	0.0621054
16	4.28E−05	0.00031296	0.00324406
32	0.00255137	0.00613996	0.01612947
64	0.36120341	0.64115893	0.83964767
128	1.71383703	2.18576009	2.69262304

Table 8 Experimental results with GA for spherical function (F1)

F1			
Variables	Best	Average	Worst
8	8.66E−07	0.00094	0.0070
16	4.09E−06	0.00086	0.0083
32	1.14E−06	0.00094	0.0056
64	1.00E−05	0.00098	0.0119
128	1.00E−05	9.42E−04	0.0071

Table 9 Experimental results with GA for Rosenbrock function (F2)

F2			
Variables	Best	Average	Worst
8	5.29E−05	0.05823	0.30973
16	0.00071	0.05683	0.50171
32	0.00228	0.05371	0.53997
64	0.00055	0.053713	0.26777
128	0.000286	0.05105	0.26343

Table 10 Experimental results with GA for Ackley function (F3)

F3			
Variables	Best	Average	Worst
8	3.006976	3.14677173	3.38354
16	3.163963	3.351902975	3.57399568
32	3.246497	3.14677173	3.86201
64	3.519591	3.86961452	4.15382873
128	3.8601773	4.209902992	4.55839099

Table 11 Experimental results with GA for Rastringin function (F4)

F4

Variables	Best	Average	Worst
8	0.499336	6.7430	15.3442
16	8.160601	24.01	43.39
32	46.008504292	82.35724	129.548
64	162.4343	247.0152194	347.216184
128	524.78094	672.6994	890.93943

Table 12 Experimental results with GA for Griewank function (F5)

F5

Variables	Best	Average	Worst
8	0.001547453	0.026897950	0.09962
16	0.0053780643	0.12157227	0.34964841
32	0.141923311	0.410196991	0.9173677
64	0.7874362847	0.980005731	1.00242183
128	1.0051894441	1.006888465	1.00810391

variables, we can find the best, average and worst results with each of the mathematical functions.

The results in Table 3 show that the CS with static parameters 50 experiments for Spherical Function (F1), we can find that for 8, 16, 32, 64 variables, CS showed better results compared to the Table 8 of GA.

In Table 4 we can find that CS achieve better resultsin 8 variables for Rosenbrock Function, and in Table 9 with GA achieve the best results with 16, 32, 64 and 128 variables.

CS for Ackley Function (F3) we can find in Table 5 that the algorithm surpassed the results shown in Table 10 with GA.

Table 6 achieve better results CS in 8 and 16 variables in average relative to the Table 11 the GA.

We can achieve in the Table 7, CS algorithm findbetter results for 16, 32, 64 variables, compared with Table 12 the GA.

5.2 Simulation Results with The Genetic Algorithm (GA)

In this Section we find the tables of results obtained by Genetic Algorithm for each mathematical function [16].

Table 8 find the results of GA with Spherical Function (F1), Unlike the results find in Table 3 the CS, GA shows better results for 128 variables.

We can see in Table 9 that GA, find better results as opposed to the results of the CS in Table 4 for 16, 32, 64 and 128 variables.

In comparing the results of Table 5 the CS, Table 10 the GA not exceeded results the CS.

In the Table 11 the GA has better average for variable 32, 64 and 128, in comparison with the results of CS in Table 6.

Table 12 The GA, find good results in average for the variables 8 and 128, in comparison with the Table 7 presents good results for 16, 32, 64 variables.

6 Conclusions

In the results of the CS algorithm, we can analyze it according to the parameters selected for this algorithm there may be variations in results, allowing the method to converge faster, we can denote that good results are obtained compared to the genetic algorithm, but, we have to work harder to find out what are the parameters that affect the CS algorithm and improve on Rosenbrock and Rastringin functions to gradually increase the number of variables.

The CS algorithm has many variants, and has been applied to programming problems nursing, training neural networks, in software testing and test data, generation in wireless sensor networks, among other applications, has shown promising results CS.

Acknowledgment We thank CONACYT and Tijuana Institute of Technology for the facilities and resources granted for the development of this research.

References

1. Chifu, V.R., Pop, C.B., Salomie, I., Suia, D.S., Niculici, A.N.: Optimizing the semantic web service composition process using cuckoo search. In: Intelligent Distributed Computing V. Studies in Computational Intelligence, vol. 382, pp. 93–102 (2012)
2. Bhargava, V., Fateen, S.E.K., Bonilla-Petriciole, A.: Cuckoo Search: A New Nature-Inspired Optimization Method for Phase Equilibrium Calculations, vol. 337, pp. 191–200 (2013). doi:10.1016/j.fluid.2012.09.018
3. Choudhary, K., Purohit, G.N.: A new testing approach using cuckoo search to achieve multi-objective genetic algorithm. J. Comput. pp. 117–119 (2001)
4. Dhivya, M., Sundarambal, M., Anand, L.N.: Energy efficient computation of data fusion in wireless sensor networks using cuckoo based particle approach (CBPA). Int. J. Commun. Netw. Syst. Sci. **4**, 249–255 (2001)
5. Dhivya, M., Sundarambal, M.: Cuckoo search for data gathering in wireless sensor networks. Int. J. Mobile Commun. **9**, 642–656 (2011)
6. Eiben, A.E., Smith, J.E.: Introduction to Evolutionary Computing (Natural Computing Series). Springer, Berlin (2013)
7. Gandomi, A.H., Yang, X.S., Alavi, A.H.: Cuckoo search algorithm: a meteheuristic approach to solve structural optimization problems. Eng. Comput. **29**(1), 17–35 (2013)

8. Gandomi, A.H., Yang, X.S., Talatahari, S., Deb, S.: Coupled eagle strategy and differential evolution for unconstrained and constrained global optimization. Comput. Math. Appl. **63**, 191–200 (2012)
9. Jamil, M., Yang, X.-S.: A literature survey of benchmark functions for global optimization problems. Int. J. Math. Model. Numer. Optim. **4**, 150–194 (2013)
10. Jamil, M., Zepernick, H.: Multimodal function optimisation with cuckoo search algorithm. Int. J. Bio-inspired Comput. **5**, 73–83 (2013)
11. Ong, P., Zainuddin, Z.: An efficient cuckoo search algorithm for numerical function optimization, In: AIP Conference Proceedings, vol. 1522, pp. 1378 (2013)
12. Perumal, K., Ungati, J.M., Kumar, G., Jain, N., Gaurav, R., Srivastava, P.R.: Test data generation: a hybrid approach using cuckoo and tabu search. In: Swarm, Evolutionary, and Memetic Computing (SEMCCO2011). Lecture Notes in Computer Sciences, vol. 7077, pp: 46–54 (2013)
13. Rajabioun, R.: Cuckoo optimization algorithm. Appl. Soft Comput. **11**, 5508–5518 (2011)
14. Srivastava, P.R., Chis, M., Deb, S., Yang, X.S.: An efficient optimization algorithm for structural software testing. Int. J. Artif. Intell. **9**, 68–77 (2012)
15. Tein, L.H., Ramli, R.: Recent advancements of nurse scheduling models and a potential path. In: Proceedings of 6th IMT-GT Conference on Mathematics, Statistics and Its Applications (ICMSA 2010), pp. 395–409 (2010)
16. Valdez, F., Melin, P., Castillo, O.: Fuzzy control of parameters to dynamically adapt the PSO and GA algorithms. In: Fuzzy Systems (FUZZ), 2010 IEEE International Conference, pp. 1–8, 23 July 2010
17. Valian, E., Mohanna, S., Tavakoli, S.: Improved cuckoo search algorithm for feedforward neural network training. Int. J. Artif. Intell. Appl. **2**(3), 36–43 (2011)
18. Vazquez, R.A.: Training spiking neural models using cuckoo search algorithm. In: 2011 IEEE Congress on Evolutionary Computation (CEC'11), pp. 679–686 (2011)
19. Yang, X.-S.: Cuckoo Search and Firefly Algorithm, Theory and Applications. Springer, Heidelberg (2014)
20. Yang, X.-S., Deb, S.: Cuckoo search via Lévy flights. In: World Congress on Nature and Biologically Inspired Computing, 2009 (NaBIC 2009), pp. 210–214 (2009)
21. Yang, X.-S., Deb, S.: Engineering optimisation by cuckoo search. Int. J. Math. Model. Numer. Optim. **1**(4), 330–343 (2010)
22. Yang, X.S., Deb, S.: Multi-objective cuckoo search for design optimization. Comput. Oper. Res. **40**(6), 1616–1624 (2013)
23. Zheng, H.Q., Zhou, Y.: A novel cuckoo search optimization algorithm based on Gauss distribution. J. Comput. Inform. Syst. **8**, 420–4193 (2012)

A Harmony Search Algorithm Comparison with Genetic Algorithms

Cinthia Peraza, Fevrier Valdez and Oscar Castillo

Abstract We describe in this paper a Harmony Search (HS) Algorithm and their areas of application, variants and comparison with other existing algorithms. HS is a metaheuristic music inspired algorithm used to solve a wide range of optimization problems applied to different areas, which has been very successful as indicated by the literature. A comparison with genetic algorithms was performed to evaluate the advantages of HS.

Keywords Harmony search · Optimization problems · Mathematical functions · Genetic algorithms

1 Introduction

We describe in this paper a harmony search algorithm which is metaheuristic algorithm inspired by music. In particular we refer to the improvisation of jazz version of Hs and its comparison with the genetic algorithm. These algorithms were applied to benchmark mathematical functions and comparative tables were made showing the optimization of results between Genetic Algorithm and Harmony Search algorithm.

The comparative study of the two algorithms is performed in order to show the effectiveness of harmony search algorithm versus optimization problem, in the same manner proving that it is more effective than the genetic algorithm.

The paper is organized as follows: in this Sect. 2 a description about Harmony Search Algorithm is presented, in this Sect. 3 a description of Genetic Algorithm is shown, in Sect. 4 description the mathematical functions is presented, in Sect. 5 a description about the optimization problems is shown, in Sect. 6 the simulations results are described and we can appreciate a comparison between Harmony Search

C. Peraza · F. Valdez (✉) · O. Castillo
Tijuana Institute of Technology, Tijuana, BC, Mexico
e-mail: fevrier@tectijuana.mx

© Springer International Publishing Switzerland 2015
O. Castillo and P. Melin (eds.), *Fuzzy Logic Augmentation of Nature-Inspired Optimization Metaheuristics*, Studies in Computational Intelligence 574,
DOI 10.1007/978-3-319-10960-2_7

algorithm and genetic algorithms, and in Sect. 7 the conclusions obtained after the study of the two algorithms versus mathematical functions is presented.

In the literature there are works where the Harmony Search has been used. In [4] a new Meta heuristic algorithm for continuous engineering optimization Theory and practice is presented. In this paper the authors propose a new harmony search(HS) meta heuristic algorithm based approach for engineering optimization problems with continuous design variables it uses a stochastic random search instead of a gradient search so that derivative information is unnecessary various engineering optimization problems, including mathematical function minimization and structural engineering optimization problems, are presented to demonstrate the effectiveness and robustness of the HS algorithm. The results indicate that the proposed approach is a powerful search and optimization technique that may yield better solutions to engineering problems than those obtained using current algorithms. In [5] the Parameter setting free harmony search algorithm is presented, the authors proposed this study a novel technique to eliminate tedious and experience requiring parameter assigning efforts. The new parameter setting free (PSF) technique which this study suggests contains one additional matrix which contains an operation type (random selection, memory consideration, or pitch adjustment) for every variable in harmony memory. In [15] the Harmony Search Benchmarking of heuristic optimization methods is presented, the authors propose it is short history when many heuristic optimization methods appear. As example Particle swarm optimization method (PSO) or Repulsive particle swarm optimization method (RPSO), Gravitational search algorithm (GSA), Central force optimization (CFO), Harmony search algorithm (HAS) etc. Those methods are working differently but all of them can optimize same problems. There is general question: Exists any standard benchmark which can be used for individual methods comparing. It is a bit hard to answer this question because it is possible to find some optimization problems which are widely used along some papers but in fact there does not exists summary which can be uses for standard evaluation of optimization process. In [9] the Global Best Harmony Search is presented, the authors propose a new variant of HS concepts from swarm intelligence are borrowed to enhance the performance of HS. The performance of the GHS is investigated and compared with HS and a recently developed variation of HS. The experiments performed show that the GHS generally outperformed the other approach when applied to ten benchmark problems. The effect of noise on the performance of the three HS variants is investigated and a scalability study is conducted. The effect of the GHS parameters is analyzed. Finally, the three HS variants are compared on several Integer Programming test problem. The results show that the three approaches seem to be an efficient alternative for solving Integer Programming problem. In [16] the Self adaptive: harmony search algorithm for optimization is presented, the authors proposed a new metaheuristic optimization algorithm harmony search (HS) with continuous design variables was developed. This algorithm is conceptualized using the musical improvisation process of searching for a perfect state of harmony. Although several variants and an increasing number of applications have appeared, one of its main difficulties is how to select suitable parameter values. In [10] the An improved harmony search algorithm for solving optimization problems is presented, in this paper the authors propose develops an improved harmony search (IHS) algorithm for solving optimization problems. IHS employs a novel method for generating new solution vectors that enhances accuracy and convergence rate of harmony search (HS) algorithm, in [11] the a survey on applications of the harmony search algorithm, in this paper they propose thoroughly reviews and analyses the main characteristics and application portfolio of the so called Harmony Search algorithm a meta heuristic approach that has been shown to achieve excellent results in a wide range of optimization problems. In [8] the A Tabu Harmony Search Based Approach to Fuzzy Linear Regression is presented, the authors propose an unconstrained global continuous optimization method based on tabu search and harmony search to support the design of fuzzy linear regression (FLR) models. Tabu and harmony search strategies are used for diversification and intensification of FLR, respectively. The authors propose approach offers the flexibility to use any kind of an objective function based on client's requirements or requests and the nature of the data set and then attains its minimum error. In [14] the A new gravitational search algorithm using fuzzy logic to parameter adaptation the authors propose a new method using fuzzy logic to change alpha parameter and give a different gravitation and acceleration to each agent in order to improve its performance, we use this new approach for mathematical functions and present a comparison with original approach. In [3] the Fuzzy Control of Parameters to Dynamically Adapt the PSO and GA Algorithms the authors propose a new hybrid approach for mathematical function optimization combining Particle Swarm Optimization (PSO) and Genetic Algorithms (GAs) using Fuzzy Logic for parameter adaptation and integrate the results. In [7] the Music Inspired Harmony Search Algorithm Theory and practice, the authors propose we show the performance of the algorithm and the areas in which it can be applied. In [6] the Harmony Search Algorithms for structural design optimization the authors propose a show us the type of problems you can solve the harmony search algorithm and some methods that have been proposed to improve in certain areas of application. In [12] the Differential evolution with dynamic adaptation of parameters for the optimization of fuzzy controllers is presented, the authors propose a new algorithm using fuzzy logic with dynamic adaptation of parameters.

2 Harmony Search Algorithm

Harmony search is a relatively new heuristic optimization algorithm inspired music and was first developed by ZW Gemm et al. in 2001 [7].

This algorithm can be explained more in detail with the process of improvisation that takes a musician, which consists of three options:

1. Play any song you have in your memory
2. Play a similar composition to an existing
3. Play a new song or randomly

If we formalize these three options for optimization, we have three corresponding components: memory usage of harmony, pitch adjustment and randomization [17].

2.1 Memory in Harmony Search Algorithm

The use of harmony memory is important because it is similar to choosing the best people in genetic algorithms. This will ensure the best harmonies will be transferred to the new memory harmony. In order to use this memory more effectively, we can assign a parameter r_{accept} € [0, 1] call acceptance rate memory. If this rate is too low, just select the best harmonies and may converge very slowly [17].

$$r_{accept} \in [0, 1] \tag{1}$$

2.2 Pitch Adjustment

To adjust the pitch slightly in the second component, we have to use such a method can adjust the frequency efficiently. In theory, the tone can be adjusted linearly or nonlinearly, but in practice the linear is used. If the current solution is X_{old} (or pitch), then the new solution (tone) is generated X_{new}.

$$x_{new} = x_{old} + b_p(2rand - 1) \tag{2}$$

where "*rand*" is a random number drawn from a uniform distribution [0, 1]. Here is it bandwidth, which controls the local range of tone adjustment in fact, we can see that the pitch adjustment (2) is a random step.

Pitch setting is similar to the mutation operator in genetic algorithms. We can assign a pitch adjustment rate to control the degree of adjustment. If too low, there is usually no change. If too high, then the algorithm may not converge at all [17].

2.3 Randomization

The third component is a randomization component (3) that is used to increase the diversity of the solutions. Although the tone setting has a similar role, but it is limited to certain local tone adjustment and therefore correspond to a local search. The use of randomization can further push the system to explore various regions with high diversity solution in order to find the global optimum [17]. So we have:

$$P_a = P_{lower\,limit} + P_{range} * rand \tag{3}$$

where rand is a generator of random numbers in the range of 0 and 1. (Search space) $P_{range} = P_{upper\ limit} - P_{lower\ limit}$

The three components in harmony search can be summarized in the pseudo code shown in Sect. 2.4, where you can find that the probability of a true randomization (4) is

$$P_{random} = 1 - r_{accept} \tag{4}$$

And the actual probability of tone adjustment (5) is

$$P_{tono} = r_{accept} * r_{pa} \tag{5}$$

2.4 Pseudo Code for Harmony Search Algorithm

The pseudo code for HS is presented below:

Objective function f (x), x = (x₁, ...,xₙ)ᵀ
Initial generate harmonics (matrices of real numbers)
Define pitch adjustment rate (rpa) and limits of tone
Define acceptance rate of the harmony memory (r accept)
while (t <Maximum number of iterations)
Generate a new harmony and accept the best harmonies
Setting the tone for new harmonies (solutions)
if (rand>raccept)
Choose an existing harmony randomly
else if (rand>rpa)
Setting the tone at random within a bandwidth (2)
else
Generate a new harmony through a randomization (3)
End if
Accepting new harmonies (solutions) best
End while
To find the best solutions.

2.5 Variants

There are three variants of the algorithm that have been applied to achieve better results briefly explain each of them:

The improved harmony search algorithm (IHS)

To address the shortcomings of the HS, Mahdavi et al. [10] proposed a new variant of the HS, called the improved harmony search (IHS). The IHS dynamically updates (r_{pa}) according to the following equation,

$$Rpa(t) = Rpa_{\min} + \frac{(Rpa_{\max} - Rpa_{\min})}{NI} + t \tag{6}$$

where $Rpa(t)$ is the pitch adjusting rate for generation t, $PAR_{\min}$ is the minimum adjusting rate, $PAR_{\max}$ is the maximum adjusting rate and t is the generation number.

In addition, bp is dynamically updated as follows:

$$bp(t) = bp_{\max}{}^e\left(\frac{\ln\left(\frac{bp_{\min}}{bp_{\max}}\right)}{NI}\right) * t \tag{7}$$

where $bp(t)$ is the bandwidth for generation t, $bp_{\min}$ is the minimum bandwidth and $bp_{\max}$ is the maximum bandwidth.

A major drawback of the IHS is that the user needs to specify the values for $bw_{\min}$ and $bp_{\max}$ which are difficult to guess and problem dependent.

Best overall harmony search (GHS)

Inspired by the concept of swarm intelligence as proposed in Particle Swarm Optimization (PSO) [2], a new variation of HS is proposed in this paper. In a global best PSO system, a swarm of individuals (called particles) fly through the search space. Each particle represents a candidate solution to the optimization problem.

The position of a particle is influenced by the best position visited by itself (i.e. its own experience) and the position of the best particle in the swarm (i.e. the experience of swarm).

The new approach, called global-best harmony search (GHS), modifies the pitch-adjustment step of the HS such that the new harmony can mimic the best harmony in the HM. Thus, replacing the bp parameter altogether and adding a social dimension to the HS. Intuitively, this modification allows the GHS to work efficiently on both continuous and discrete problems.

The GHS has exactly the same steps as the IHS with the exception that Step 3 is modified as follows:

for each i $\in$ [1,N] do
If $U(0,1) \leq R_{accept}$ then /*memory consideration */
Begin
$X_i^j = x_i^j$, where j ~ U (1,...,HMS)
If $U(0,1) \leq Rpa(t)$ then /* pitch adjustment */
Begin
$x_i^j = x_k^{best}$, where best is the index of the best harmony in the HM and k ~
U (1,N)
Endif
Else /* random selection */
$x_i^j = LB_i + r * (UB_i - LB_i)$
Endif
done

New global harmony search (NGHS)

The NGHS algorithm is an improved version of harmony search algorithm (HS), and it includes two important operations: position updating and genetic mutation with a low probability. The former can enhance the convergence of the NGHS, and the latter can effectively prevent the NGHS from trapping into the local optimum. Based on a large number of experiments, the NGHS has demonstrated high efficiency on solving chemical equation balancing. The results show that the NGHS can be an efficient alternative for solving chemical equation balancing [1].

2.6 Application Areas

The harmony search algorithm has been applied so far to various optimization problems. Moreover, the structure of the algorithm has been customized by case adjust basic structure. To overcome this situation, the algorithm of harmony search (HS) used a new stochastic derivative, using the experiences of musicians in jazz improvisation and may be applicable to discrete variables. Instead of tilting the information of an objective function, the stochastic derivative HS gives a probability of being selected for each value of a decision variable.

The HS algorithm has been applied to various problems in science and engineering optimization including:

Optimization function, the distribution of water, groundwater modeling, energy saving clearance, structural design, vehicle routing, and others. The possibility of combining harmony search with other algorithms such as particle swarm optimization and genetic algorithms has also been investigated [17].

3 Genetic Algorithms

Genetic algorithms (GA) emulate genetic evolution. The characteristics of individuals are therefore expressed using genotypes. The original form of the GA, as illustrated by John Holland in 1975, had the distinct features: (1) a bit string representation, (2) proportional selection, and (3) cross-over as the primary method to produce new individuals. Since then, several variations to the original Holland GA have been developed, using different representation schemes, selection, crossover, mutation and elitism operators [2].

3.1 Representation

The classical representation scheme for GAs is of binary vectors of fixed length. In the case of an n_x-dimensional search space, each individual consists on n_x variables with each variable encoded as a bit string. If variables have binary values, the length of each chromosome is n_x bits. In the case of nominal-valued variables, each nominal value can be encoded as an n_d-dimensional bit vectors where 2nd is the total numbers of discrete nominal values for that variable. Each n_d-bit string represents a different nominal value. In the case of continuous-valued variables, each variable should be mapped to an n_d-dimensional bit vector,

$$\varphi : R \rightarrow (0, 1)^{n_d} \tag{8}$$

The range of continuous space needs to be restricted to a finite range, $[x_{min}, x_{max}]$. Using the standard binary decoding, each continuous variable x_{ij} of chromosome x_i is encoded using a fixed length bit string.

GAs have also been developed that use integer or real valued representations and order based representations where the order of variables in a chromosome plays an important role. Also, it is not necessary that chromosomes be of fixed length [2].

3.2 Crossover Operations

Several crossover operators have been developed for GA's depending on the format in which individuals are represented. For binary representations, uniform crossover, one-point crossover and two-point crossover are the most popular:

- **Uniform Crossover**, where corresponding bit positions are randomly exchanged between the two parents to produce two offspring.
- **One-Point Crossover**, where a random bit position is selected, and the bit substrings after the selected bit are swapped between the two parents to produce two offspring.

- **Two-Point Crossover**, where two bit positions are randomly selected and the bit substrings between the selected bit positions are swapped between the two parents.

For continuous valued genes, arithmetic crossover can be used:

$$x_{ij} = r_j x_{1j} + (1.0 - r_j) x_{2j} \tag{9}$$

where $r_j \sim U(0, 1)$ and x_i is the offspring produced from parents x_1 and x_2 [2].

3.3 Mutation

The mutation scheme used in a GA depends on the representation scheme. In the case of bit string representations, we here:

- **Random Mutation**, randomly negates bits, while
- **In-Order Mutation**, performs random mutation between two randomly selected bit positions.

For discrete genes with more than two possible values that a gene can.

Assume, random mutation selects a random value from the finite domain of the gene. In the case of continuous valued genes, a random value sampled from a Gaussian distribution with zero mean and small deviation is usually added to the current gene value. As an alternative, random noise can be sampled from a Cauchy distribution [2].

4 Benchmark Mathematical Functions

This section list a number of the classical benchmark functions used to validate optimization algorithms.

In the area of optimization using mathematical functions have been considered in the works mentioned below: A new gravitational search algorithm using fuzzy logic to parameter adaptation [14], Differential evolution with dynamic adaptation of parameters for the optimization of fuzzy controllers [12], Bat algorithm comparison with genetic algorithm using benchmark functions [13].

To validate our method we used a set of 6 benchmark mathematical functions, called Spherical, Rosenbrock, Rastrigin, Ackley, Zakharov, Sum Square; all functions were evaluated with 4, 5, 10, 20, 30 and 40 Harmonies.

Figure 1 shows the plot corresponding to the Spherical function and Eq. 10 represents the Spherical function. Figure 2 shows the plot corresponding to Rosenbrock function and Eq. 11 represents the Rosenbrock function.

Fig. 1 Spherical function

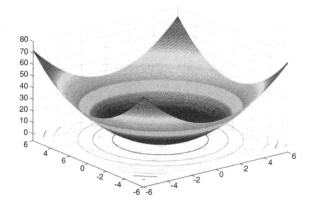

Fig. 2 Rosenbrock function

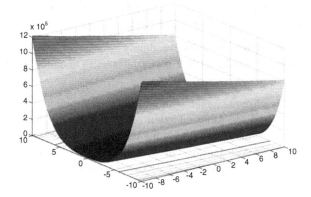

The mathematical functions are shown below:

$$f(x) = \sum_{j=1}^{n_x} x_j^2 \tag{10}$$

Witch $x_j \in [-100, 100]$ and $f^*(x) = 0.0$

$$f(x) = \sum_{j=1}^{n_z/2} [100(x_{2j} - x_{2j-1}^2)^2 + (1 - x_{2j-1})^2] \tag{11}$$

Witch $x_j \in [-2.048, 2.048]$ and $f^*(x) = 0.0$

Figure 3 shows the plot corresponding to Rastrigin function and Eq. 12 shows the description the Rastrigin function.

$$f(x) = \sum_{j=1}^{n_x} (x_j^2 - 10\cos(2\pi x_j) + 10) \tag{12}$$

With $x_j \in [-5.12, 5.12]$ and $f^*(x) = 0.0$

Fig. 3 Rastrigin function

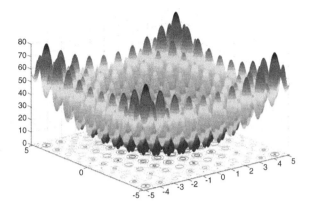

Fig. 4 Ackley function

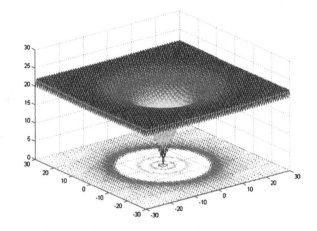

Figure 4 shows the plot corresponding to Ackley function and Eq. 13 shows the description the Ackley function.

$$f(x) = -20e^{-0.2\sqrt{\frac{1}{n_x}\sum_{j=1}^{n_x} x_j^2 - \frac{1}{n_x}\sum_{j=1}^{n_x}\cos(2\pi x_j)}} + 20 + e \tag{13}$$

With $x_j \in [-30, 30]$ and $f^*(x) = 0.0$

Figure 5 shows the plot corresponding to Zakharov function and Eq. 14 shows the description the Zakharov function.

$$f(x) = \sum_{i=1}^{n} x_i^2 + (\sum_{i=1}^{n} 0.5ix_i)^2 + (\sum_{i=1}^{n} 0.5ix_i)^4 \tag{14}$$

Witch $x_i \in [-5, 10]$ and $f^*(x) = 0.0$

Fig. 5 Zakharov function

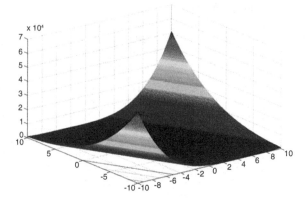

Fig. 6 Sum square function

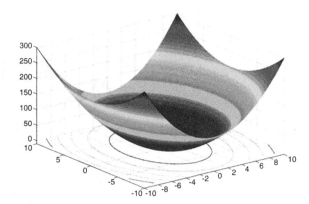

Figure 6 shows the plot corresponding to Sum Square function and Eq. 15 shows the description the Sum Square function.

$$f(x) = \sum_{i=1}^{n} i x_i^2 \qquad (15)$$

Witch $x_i \in [-2, 2]$ and $f^*(x) = 0.0$

5 Optimization Problems

The optimization problem can be defined as:

$$\text{Min } y = f(x) \qquad (16)$$

Basically the problems can be divided into unimodal and multimodal.

Another aspect are constrains. Some problems have not constrains. The best optimization method has to find optimal value in all cases.

There exist many algorithms for problem solving. Some of them are special, some are more general. Many problems cannot be solved by deterministic algorithm, so heuristic algorithm are used. In the set of metaheuristic algorithms we can find PSO [3], GSA [14], DE [12] and others.

Unimodal functions

Have only one local optimum. Those functions are relatively easy to analyze for optimums. They are used for checking the speed of optimization and convergence. There are used commonly two functions. First one is Schefel's and second one is Rastrigin's.

Multimodal functions

Multimodal functions have multiple local optimums. Some methods will stuck in local optimum. Main goal of those problems is to test solvers how they are able to avoid local optimum. Some problems have no single global optimum. Some of them have one global optimum and many local which are very close in the term of fitness function. First one is Sphere, second one is Sum and Product and third one is Griewank's.

6 Simulation Results

In this section the comparison of the Harmony Search algorithm is made against genetic algorithms [13]. In each of the algorithms 6 mathematical functions Benchmark were considered separately, a dimension of 10 variables was used with 30 runs for each function varying the parameters of the algorithms.

The parameters used in the HS were:

- Size solution harmonies: 4–40 Harmonies.
- Harmony memory accepting: 0.75–0.95.
- Pitch adjustment: 0.1–0.5.
- Pitch range: 200–400.

The parameters for the genetic algorithm are shown below [13]:

- Number of Individuals: 4–40.
- Selection: Stochastic, Remainder, Uniform, Roulette.
- Crossover: Scattered, Single Point, Two Point, Heuristic and Arithmetic.
- Mutation: Gaussian, Uniform.

6.1 Simulation Results with Harmony Search Algorithm

In this section we show the experimental results obtained by the Harmony Search algorithm in separate tables of the mathematical functions. Table 1 shows the simulation results for the Spherical function.

From Table 1 it can be appreciated that after executing the HS Algorithm 30 runs, with different parameters, we can find the best, average and worst results for the Spherical function. Table 2 shows the simulation results for the Rosenbrock function.

From Table 2 it can be appreciated that after executing the HS Algorithm 30 runs, with different parameters, we can find the best, average and worst results for the Rosenbrock function. Table 3 shows the simulation results for the Rastrigin function.

Table 1 Simulation results for the spherical function

Number of harmonies	Best	Worst	Mean
4	0.000038975	0.00014863	0.0000762868
5	0.000039501	0.00014258	0.0000792732
10	0.000053865	0.00012135	0.0000883132
20	0.000029966	0.000079598	0.0000476648
30	0.000024495	0.000080963	0.0000499808
40	0.000023567	0.000077813	0.0000523875

Table 2 Simulation results for the Rosenbrock function

Number of harmonies	Best	Worst	Mean
4	0.0000000010631	0.00000039836	0.0000000578908
5	0.00000000073035	0.00000032337	0.0000000521
10	0.000000000079568	0.00000023364	0.0000000475
20	0.000000000014716	0.001	0.0000387
30	0.0000000010014	0.001	0.00112
40	0.00000000010979	0.0079	0.00176

Table 3 Simulation results for the Rastrigin function

Number of harmonies	Best	Worst	Mean
4	0.000000000011045	0.00000007236	0.0000000139
5	0.0000000000093081	0.000000011282	0.0000000126
10	0.00000000061454	0.00000022195	0.0000000324
20	0.000000000032507	0.000000092489	0.0000000172
30	0.00000000006964	0.00000022386	0.0000000321
40	0.0000000037211	0.00000017014	0.0000000275

Table 4 Simulation results for the Ackley function

Number of harmonies	Best	Worst	Mean
4	0.0000098012	0.00031775	0.000122722
5	0.0000087238	0.00017561	0.0000969749
10	0.000022005	0.00035406	0.000139821
20	0.0000093788	0.00028147	0.0000940209
30	0.000016206	0.00040138	0.000136185
40	0.000031467	0.00045383	0.000116993

Table 5 Simulation results for the Zakharov function

Number of Harmonies	Best	Worst	Mean
4	0.000095054	0.00032929	0.00019524
5	0.00007544	0.00033906	0.00022226
10	0.000095173	0.00040378	0.00027224
20	0.000073328	0.00037254	0.00019199
30	0.000077163	0.00036816	0.00022529
40	0.00019508	0.0018	0.00052135

From Table 3 it can be appreciated that after executing the HS Algorithm 30 runs, with different parameters, we can find the best, average and worst results for the Rastrigin function. Table 4 shows the simulation results for the Ackley function.

From Table 4 it can be appreciated that after executing the HS Algorithm 30 runs, with different parameters, we can find the best, average and worst results for the Ackley function. Table 5 shows the simulation results for the Zakharov function.

From Table 5 it can be appreciated that after executing the HS Algorithm 30 runs, with different parameters, we can find the best, average and worst results for the Zakharov function. Table 6 shows the simulation results for the Sum Square function.

Table 6 Simulation results for the sum square function

Number of harmonies	Best	Worst	Mean
4	0.0000086056	0.000037648	0.00002513
5	0.0000065348	0.000041995	0.0000238824
10	0.000021245	0.00010459	0.000064354
20	0.0000048182	0.000046886	0.000029147
30	0.000011206	0.000054988	0.0000351173
40	0.000017024	0.000056469	0.0000379489

From Table 6 it can be appreciated that after executing the HS Algorithm 30 runs, with different parameters, we can find the best, average and worst results for the Sum Square function.

6.2 Simulation Results with the Genetic Algorithm

In this section we show the experimental obtained by the genetic algorithm in separate tables of the mathematical functions [13]. Table 7 shows the simulation results for the Sum Sphere function using genetic algorithm.

From Table 7 it can be appreciated that after executing the Genetic Algorithm 30 runs, with different parameters, we can find the best, average and worst results for the Sphere function. Table 8 shows the simulation results for the Rosenbrock function using genetic algorithm.

From Table 8 it can be appreciated that after executing the Genetic Algorithm 30 runs, with different parameters, we can find the best, average and worst results for the Rosenbrock function. Table 9 shows the simulation results for the Rastrigin function using genetic algorithm.

From Table 9 it can be appreciated that after executing the Genetic Algorithm 30 runs, with different parameters, we can find the best, average and worst results

Table 7 Simulation results for the sphere function

Population	Best	Worst	Mean
4	0.000655746	0.867154805	0.501445118
5	0.016158419	0.735175081	0.297672568
10	0.029477858	0.985900891	0.616489628
20	0.125851757	1.018067545	0.250640697
30	0.050431819	0.928690136	0.521845011
40	0.004944109	1.847289399	0.558953430

Table 8 Simulation results for the Rosenbrock function

Population	Best	Worst	Mean
4	0.089847334	0.802024183	0.561886727
5	0.045156568	0.878087097	0.476680695
10	0.026476082	0.788665597	0.212878969
20	0.008755795	0.654965394	0.151633433
30	0.001220403	0.292128413	0.050677876
40	0.000245092	0.843183891	0.242982579

Table 9 Simulation results for the Rastrigin function

Population	Best	Worst	Mean
4	0.014939893	0.997649164	0.532383503
5	0.025389969	1.023785562	0.484650675
10	0.000983143	1.991943912	0.591455815
20	0.005860446	0.973098541	0.416048173
30	0.001461684	1.030270667	0.402938563
40	0.0017108	1.011176844	0.304818043

for the Rastrigin function. Table 10 shows the simulation results for the Ackley function using genetic algorithm.

From Table 10 it can be appreciated that after executing the Genetic Algorithm 30 runs, with different parameters, we can find the best, average and worst results for the Ackley function. Table 11 shows the simulation results for the Zakharov function using genetic algorithm.

From Table 11 it can be appreciated that after executing the Genetic Algorithm 30 runs, with different parameters, we can find the best, average and worst results for the Zakharov function. Table 12 shows the simulation results for the Sum Square function using genetic algorithm.

Table 10 Simulation results for the Ackley function

Population	Best	Worst	Mean
4	0.003764969	0.974296237	0.346336685
5	0.034816548	1.008082845	0.555913584
10	0.00068874	1.006617458	0.309683405
20	0.00000943	0.999864848	0.217317909
30	0.0024992	0.045123886	0.01309672
40	0.00110442	0.963070253	0.229423519

Table 11 Simulation results for the Zakharov function

Population	Best	Worst	Mean
4	0.00291835	0.974030463	0.442892629
5	0.003929128	2.494950163	0.368494675
10	0.000621031	0.999709009	0.475767117
20	0.00754580	0.787983798	0.444793802
30	0.00074412	4.568706165	1.518534814
40	0.01262987	2.859991576	1.30142878

Table 12 Simulation results for the sum square function

Population	Best	Worst	Mean
4	0.029476905	0.834760272	0.379557704
5	0.009588642	0.441979109	0.172141341
10	0.000848578	0.045665911	−0.002597231
20	0.01018758	0.668265573	0.194550999
30	0.00259567	0.226922795	0.103921866
40	0.00638302	0.406954403	0.149111035

From Table 12 it can be appreciated that after executing the Genetic Algorithm 30 runs, with different parameters, we can find the best, average and worst results for the Sum Square function.

7 Conclusions

The HS algorithm is a new method which can solve various types of problem very easily and effectively because it not requires many complex calculations. The HS can handle discrete, continuous variables and can be applied to linear and nonlinear functions.

In the analysis of results obtained with the genetic algorithm and harmony search, we conclude that the HS is better than the GA this is demonstrated with the tables mentioned in the previous section to obtain a minimum error in all the benchmark functions which was applied, the same number of dimensions were used to perform the comparison.

The analysis of simulation results between HS and GA method considered in this work, lead us to the conclusion that for the optimization of benchmark functions, the HS method is a good alternative because it is easier to optimize and achieve good results try that with GA.

As we can realize in each of the tables where the results of the experiments with HS is best values were obtained.

With this it has been that the algorithm HS is better than GA.

Acknowledgments We would like to express our gratitude to the CONACYT and Tijuana Institute of Technology for the facilities and resources granted for the development of this research.

References

1. Dexuan, Z., Yanfeng, G., Liqun, G., Peifeng, W.: A novel global harmony search algorithm for chemical equation balancing. International Conference on Computer Design and Appliations, pp. 1–3. IEEE (2010)

2. Eberhart, R., Kennedy, J.: A new optimizer using particle swarm theory, In: Proceedings of the 6th International Symposium on Micromachine and Human Science, pp. 39–43. IEEE (1995)
3. Fevrier, V., Melin, P., Oscar, C.: Fuzzy Control of Parameters to Dynamically Adapt the PSO and GA Algorithms, pp. 1–8. IEEE, Barcelona, Spain (2010)
4. Geem, Z., Lee, K.: A New Meta-Heuristic Algorithm for Continuous Engineering Optimization Harmony Search Theory and Practice, Department of Civil and Environmental Engineering, University of Maryland, College Park, pp. 3-20. Elsevier, Maryland, USA (2004)
5. Geem, Z., Sim, K.: Parameter Setting Free Harmony Search Algorithm, School of Electrical and Electronics Engineering, Chung Ang University, pp. 2–10. Elsevier, Chung Ang, China (2010)
6. Geem, Z.: Harmony Search Algorithms For Structural Design Optimization. Studies in Computational Intelligence, pp. 8–121. Springer, Heidelberg, Germany (2009)
7. Geem Z.: Music Inspired Harmony Search Algorithm Theory and Applications, Studies in Computational Intelligence, pp. 8–121, Springer, Heidelberg, Germany (2009)
8. Hadi, M., Mehmet, A., Mashinchi, M., Pedrycz, W.:A Tabu Harmony Search Based Approach to Fuzzy Linear Regression, Transactions on Fuzzy Systems, pp. 1–13. IEEE, New Jersey, USA (2011)
9. Mahamed, G., Mahdavi, M.: Global Best Harmony Search, Applied Mathematics and Computation, pp. 1–14. Elsevier, Amsterdam, Holland (2008)
10. Mahdavi, M., Fesanghary, M., Damangir, E.: An Improved Harmony Search Algorithm for Solving Optimization Problems, Applied Mathematics and Computation, pp. 1567–1579. Elsevier, Amsterdam, Holland (2007)
11. Manjarres, D., Torres, L., Lopez, S., DelSer J, Bilbao M., Salcedo S., Geem Z.: A Survey on Applications of the Harmony Search Algorithm, Engineering Applications of Artificial Intelligence, pp. 3–14, Elsevier, Amsterdam, Holland (2013)
12. Ochoa, P., Castillo, O., Soria, J., Differential Evolution with Dynamic Adaptation of Parameters for the Optimization of Fuzzy Controllers, Recent Advances on Hybrid Approaches for Designing Intelligent Systems, pp. 275–288. Springer, Heidelberg, Germany (2013)
13. Perez, J., Valdez, F., Castillo, O.: Bat Algorithm Comparison with Genetic Algorithm Using Benchmark Functions, Recent Advances on Hybrid Approaches for Designing Intelligent Systems, pp. 225–237, Springer (2013)
14. Sombra, A., Valdez, F., Melin, P., Castillo, O.: A new gravitational search algorithm using fuzzy logic to parameter adaptation. IEEE Congress on Evolutionary Computation, pp. 1068–1074 (2013)
15. Štefek, A.: Benchmarking of Heuristic Optimization Methods, University of Defence, pp 1-4, IEEE, New Jersey, USA (2011)
16. Wang, C., Huang, Y.: Self Adaptive Harmony Search Algorithm for Optimization, Department of Computer Science and Information Engineering, National Yunlin University of Science and Technology, pp. 1–12, Elsevier (2010)
17. Yang, X.: Nature Inspired Metaheuristic Algorithms, 2nd edn, pp 73–76. Luniver Press, London, UK (2010)

Part II
Applications

A Gravitational Search Algorithm for Optimization of Modular Neural Networks in Pattern Recognition

Beatriz González, Fevrier Valdez, Patricia Melin
and German Prado-Arechiga

Abstract The Gravitational Search Algorithm (GSA) is a novel heuristic optimization method based on the laws of gravity and mass interactions. We described in this paper a gravitational search algorithm to optimize the architecture of the modular neural network for recognition of medical images. In this case, we are using an echocardiograms database. Results obtained with this database are good; in this case the best learning algorithm was scaled conjugate gradient (SCG) with 90.27 % recognition rate comparing with gradient descent with adaptive learning rate backpropagation (GDA) with 84.72 %.

Keywords Modular neural network · Gravitational search algorithm · Pattern recognition · Echocardiograms · GSA

1 Introduction

In the last years, the interest in algorithms inspired by natural phenomena, has grown considerably [5–8]. It has been shown by many researchers that these algorithms are well suited to solve complex problems, for example Genetic Algorithm (GA) [13], Ant Colony Search Algorithm (ACO) [3], Particle Swarm Optimization (PSO) [7], etc.

The Gravitational Search Algorithm (GSA) is a novel heuristic optimization method based on the laws of gravity and mass interactions. We describe in this paper the use of the GSA algorithm to optimize the architecture of the Modular Neural Network (MNN) for echocardiogram recognition.

This paper focuses on the field of nature inspired computation and several approaches have been studied about optimization of modular neural networks in

B. González · F. Valdez · P. Melin (✉)
Tijuana Institute of Technology, Calzada Tecnologico s/n, Tijuana, Mexico
e-mail: pmelin@tectijuana.mx

G. Prado-Arechiga
Excel Medical Center, Paseo de los Heroes 2507 Zona Rio, Tijuana, Mexico

© Springer International Publishing Switzerland 2015
O. Castillo and P. Melin (eds.), *Fuzzy Logic Augmentation of Nature-Inspired Optimization Metaheuristics*, Studies in Computational Intelligence 574,
DOI 10.1007/978-3-319-10960-2_8

pattern recognition, and some can be found in [4, 10, 14, 15]. Research in medical imaging is growing on the last years as it is normally a non-invasive method of diagnosis [1, 9]. For this reason, we are considering the application of the GSA algorithm to optimize the architecture of the modular neural network for recognition of medical images. In this case, we are using an echocardiograms database.

Normally there are many speckle noise points on the ultrasound images. So the resulting images are contaminated with this noise that corrodes the borders of the cardiac structures [2]. This characteristic turns difficult to perform image processing, and specially the pattern recognition. Besides this kind of noise, other factors influence the outcome of ultrasound image recognition.

Furthermore, the poor imaging quality of 2D echo videos due to low contrast, speckle noise, and signal dropouts, also cause problems in image interpretation [9].

The rest of the paper describes this approach in detail and is organized as follows. In Sect. 2, we describe basic concepts such as, modular neural network, the law of gravity and second motion law, gravitational search algorithm. In Sect. 3 describe the Modular Neural Network Architecture and the database of echocardiograms Recognition. In Sect. 4 experimental results se presented. In Sect. 5 the conclusions are presented.

2 Basic Concepts

2.1 Modular Neural Networks

Modular neural networks have several advantages: Each module often addresses a simpler task, and hence can be trained in fewer iterations than the monolithic neural network. Each module is small, with fewer weights than the monolithic neural network, so that the time taken for each module's training iteration. Modules can often be trained independently, in parallel.

2.2 The Law of Gravity and Second Motion Law

Isaac Newton proposed the law of gravity stating that "The gravitational force between two particles is directly proportional to the product of their masses and inversely proportional to the square of the distance between them" [11]. The gravity force is present in each object in the universe and its behavior is called "action at a distance", this means gravity acts between separated particles without any intermediary and without any delay. The gravity law is represented by the following equation:

$$F = G\frac{M_1M_2}{R^2} \tag{1}$$

where:

F	is the magnitude of the gravitational force,
G	is gravitational constant,
M_1 and M_2	are the mass of the first and second particles respectively and,
R	is the distance between the two particles.

The Gravitational search algorithm furthermore to be based on Newtonian gravity law it is also based on Newton's second motion law, which says "The acceleration of an object is directly proportional to the net force acting on it and inversely proportional to its mass" [12]. The second motion law is represented by the following equation:

$$\alpha = \frac{F}{M} \tag{2}$$

where:

a	is the magnitude of acceleration,
F	is the magnitude of the gravitational force and,
M	is the mass of the object.

2.3 Gravitational Search Algorithm

The approach was made for E. Rashedi et al., where they introduce a new algorithm for finding the best solution in problem search spaces using physical rules. Based on populations and the same time it takes as fundamental principles the law of gravity and second motion law, its principal features are that agents are considered as objects and their performance is measured by their masses, all these objects are attract each other by the gravity force, and this force causes a global movement of all objects, the masses cooperate using a direct form of communication, through gravitational force, an agent with heavy mass correspond to good solution therefore its move more slowly than lighter ones, finally its gravitational and inertial masses are determined using a fitness function [12].

We can notice that in Eq. (1) appears the gravitational constant G, this is a physic constant which determines the intensity of the gravitational force between the objects and it is defined as a very small value. The equation by which G is defined is:

$$G(t) = G(t_0)x\left(\frac{t_0}{t}\right)^{\beta}, \quad \beta < 1. \tag{3}$$

where:

$G(t)$	is the value of the gravitational constant at time t and,
$G_0(t)$	is the value of the gravitational constant at the first cosmic quantum-interval of time t_0.

The way in which the position of a number N of agents is represented by

$$X_i = (X_i^1, \ldots, X_i^d, \ldots, X_i^n) \quad \text{for } i = 1, 2, \ldots, N, \tag{4}$$

where X_i^d presents the position of ith agent in the dth dimension.

Now, Eq. (1) with new concepts of masses is defined as following: the force acting on mass i from mass j in a specific time t, is

$$F_{ij}^d(t) = G(t) \frac{M_{pi}(t) x M_{aj}}{R_{ij}(t) + \varepsilon} (x_{ij}^d(t) - d_j^d(t)) \tag{5}$$

where M_{aj} is the active gravitational mass related to agent j, M_{pi} is the passive gravitational mass related to agent I, G(t) is gravitational constant at time t, ε is a small constant, and $R_{ij}(t)$ is the Euclidian distance between two agents I and j:

$$R_{ij}(t) = \|X_i(t), X_j(t)\| \tag{6}$$

The stochastic characteristic of this algorithm is based on the idea of the total force that acts on agent i in a dimension d be a randomly weighted sum of dth components of the forces exerted from other agents,

$$F_i^d(t) = \sum_{j=i, j \neq 1} rand_i F_{ij}^d(t) \tag{7}$$

where $rand_j$ is a random number in the interval [0,1]. The acceleration now is showed as,

$$a_i^d(t) = \frac{F_i^d}{M_{ii}(t)} \tag{8}$$

where M_{ii} is the inertial mass of ith agent. For determine the velocity of an agent we considered as a fraction of its current velocity added to its acceleration.

$$V_i^d(t+1) = rand_i x V_i^d(t) + a_i^d(t) \tag{9}$$

The position of agents could be calculated as the position in a specific time t added to its velocity in a time $t + 1$ as follows,

$$x_i^d(t+1) = X_i^d(t) + V_i^d(t+1) \tag{10}$$

In this case the gravitational constant G is initialized at the beginning and will be reduced with time to control the search accuracy. Its Equation is:

$$G(t) = G(G_0, t) \qquad (11)$$

This is because G is a function of the initial value G_0 and time t. As mentioned previously gravitational and inertia masses are simply calculated by the fitness evaluation and a heavier mass means a more efficient agent. The update the gravitational and inertial masses is performed with the following equations,

$$M_{ai} = M_{pi} = M_{ii} = M_i, \quad i = 1, 2, \ldots, N, \qquad (12)$$

$$m_i(t) = \frac{fit_i(t) - worst(t)}{best_t(t) - worst(t)} \qquad (13)$$

$$M_i(t) = \frac{m_i(t)}{\sum_{j=1}^{N} m_i(t)} \qquad (14)$$

the fitness value of the agent i at time t is defined by $fiti(t)$, and $best(t)$ and $worst(t)$ are represented as:

$$best(t) = \min_{j \in \{1,\ldots,N\}} fit_j(t) \qquad (15)$$

$$worst(t) = \max_{j \in \{1,\ldots,N\}} fit_j(t) \qquad (16)$$

If we want to use GSA for a maximization problem you only have to change Eqs. (15) and (16) as following,

$$best(t) = \max_{j \in \{1,\ldots,N\}} fit_j(t) \qquad (17)$$

$$worst(t) = \min_{j \in \{1,\ldots,N\}} fit_j(t) \qquad (18)$$

The gravitational search algorithm has a kind of elitism in order that only a set of agents with bigger mass apply their force to the other. This is with objective to have a balance between exploration and exploitation with lapse of time it is achieved by the only the *Kbest* agents will attract the others *Kbest* is a function of time, with the initial value K_0 at the beginning and decreasing with time. In such a way, at the beginning, all agents apply the force, and as time passes, *Kbest* is decreased linearly and at the end there will be just one agent applying force to the others [14]. For that reason Eq. (7), can be modified as follows,

$$F_i^d(t) = \sum_{j \in Kbest, j \neq 1} rand_i F_{ij}^d(t) \tag{19}$$

where *Kbest* is the set of first K agents with the best fitness value and largest mass. A better representation of GSA process is showing it next, it is the principle of this algorithm (Fig. 1).

First an initial population is generated, next fitness of each agent is evaluated, thereafter update the gravitational constant G, *best* and *worst* of the population; next step is calculating mass and acceleration of each agent, if meeting end of iterations, in this case maximum of iterations then returns the best solution, else executes the same steps starting from fitness evaluation. Is in third step, where we apply the modification in this algorithm, we propose changing alpha parameter to update G and help to GSA a better performance.

Fig. 1 General principle of GSA. Taken of [12]

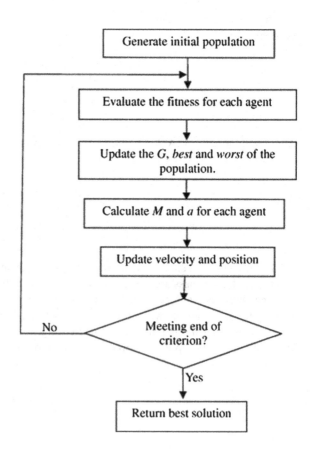

2.4 Parameters Settings in the Gravitational Search Algorithm

The following parameters were used in the setting of the GSA:

N = 20; Number of agents
max_it:10; Maximum number of iterations
$\alpha = 25$; alpha
$G_0 = 90$;

3 Modular Neural Network Architecture

We are using a Modular Neural Network in this paper. Figure 2 shows the architecture used for this work and is described as follows: We have 2 modules, where each module has one layer for input of the data, also 2 hidden layers are used and one layer in the output of the Modular Neural Network and finally, the recognize echocardiograms in the last block with this architecture.

3.1 Database of Echocardiograms Recognition

This database has the following characteristics: It contains 18 individuals, 10 images per individual. Contains images of disease patients and healthy patients. Figure 3 shows some of the echocardiograms included in this database.

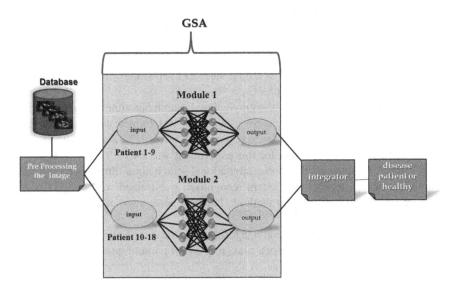

Fig. 2 Modular neural network architecture

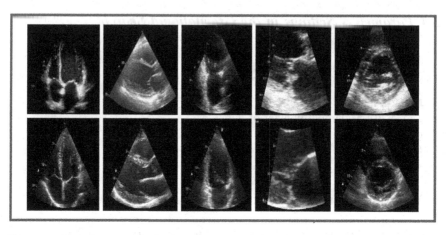

Fig. 3 Database of echocardiograms recognition

For image preprocessing, we reduced the image size from 200 × 125 to 80 × 80 pixels taken the region of interest (ROI) to eliminate as much as possible noise.

4 Experimental Results

We describe below the simulation results of our approach for echocardiograms recognition with modular neural networks (MNN). The challenge is to find the optimal architecture of this type of Modular Neural Networks, which means finding out the optimal number of layers and nodes of the neural network. We are using echocardiograms database with 10 grayscale images in bmp format of 18 subjects, 6 images by each subject were used for training the Modular Neural Network and 4 images were used to recognition. Regarding the gravitational search algorithm for Modular.

Neural Networks, we used N = 20; Number of agents, max_it = 10; Maximum number of iterations, α = 25; alpha G_0 = 90.

In Table 1 we show results of the gravitational search algorithm for optimization of modular neural networks in pattern recognition was trained with Gradient descent with adaptive learning rate backpropagation "traingda" training method. The best percentage of identification for this experiment was 84.72 %.

In Table 2 we show results of the gravitational search algorithm for optimization of modular neural networks in pattern recognition was trained with scaled Conjugate Gradient "trainscg" training method. The best percentage of identification for this experiment was 90.27 %.

Regarding the gravitational search algorithm for Modular Neural Networks also, we used N = 30; Number of agents, max_it = 20; Maximum number of iterations, α = 25; alpha G_0 = 90.

Table 1 Experimental results with Traingda training method

Epoch	Error	Training	Training time	% Ident
500	0.0001	Traingda	00:45:59	80.55
500	0.0001	Traingda	00:44:49	79.16
500	0.0001	Traingda	00:41:26	72.22
500	0.0001	Traingda	00:42:58	73.61
500	0.0001	Traingda	00:40:45	84.72
500	0.0001	Traingda	00:58:15	63.88
500	0.0001	Traingda	01:05:10	75
500	0.0001	Traingda	01:06:41	70.83
500	0.0001	Traingda	00:46:47	80.5
500	0.0001	Traingda	00:28:03	72.22

Table 2 Experimental results with Trainscg training method

Epoch	Error	Training	Training time	% Ident
500	0.0001	Trainscg	00:28:22	83.33
500	0.0001	Trainscg	00:46:15	87.5
500	0.0001	Trainscg	00:22:40	86.11
500	0.0001	Trainscg	00:28:32	87.5
500	0.0001	Trainscg	00:10:29	80.55
500	0.0001	Trainscg	00:11:58	83.33
500	0.0001	Trainscg	00:29:18	83.33
500	0.0001	Trainscg	00:30:09	80.55
500	0.0001	Trainscg	00:22:56	81.94
500	0.0001	Trainscg	00:25:08	90.27

Table 3 Experimental results with Trainscg training method

Epoch	Error	Training	Training time	% Ident
500	0.0001	Trainscg	01:02:39	70.83
500	0.0001	Trainscg	01:08:23	77.77
500	0.0001	Trainscg	01:08:26	73.61
500	0.0001	Trainscg	01:03:21	70.83
500	0.0001	Trainscg	01:04:04	80.55
500	0.0001	Trainscg	01:04:00	73.61
500	0.0001	Trainscg	01:43:25	77.77
500	0.0001	Trainscg	00:54:31	70.83
500	0.0001	Trainscg	01:02:15	77.77
500	0.0001	Trainscg	01:13:28	70.83

In Table 3 we show results of the gravitational search algorithm for optimization of modular neural networks in pattern recognition was trained with scaled Conjugate Gradient "trainscg" training method. The best percentage of identification for this experiment was 80.55 %.

5 Conclusions

The results obtained with the proposed method are good; the best learning algorithm in this case was the scaled conjugate gradient (SCG) with 90.27 %. Comparing with gradient descent with adaptive learning rate backpropagation (GDA) with 84.72 %, SCG gets the best training time and recognition. Results obtained are good, however, other methods could be used to improve results, just to make a comparison or, apply another kind of preprocessing, normally there are many speckle noise on the ultrasound images, furthermore, the poor imaging quality of echo videos due to low contrast, speckle noise, and signal dropouts also cause problems in image interpretation.

References

1. Banning, A.P., Behrenbruch, C., Kelion, A.D., Jacob, G., Noble, J.A.: A shape-space-based approach to tracking myocardial borders and quantifying regional left-ventricular function applied in echocardiography. IEEE Med. Imaging **21**, 226–238 (2002)
2. Beymer, D., Kumair, R., Tanveer, S.M., Wang, F.: Cardiac disease detection from echocardiogram using edge filtered scale-invariant motion features, Computer Vision and Pattern Recognition Workshops(CVPRW). IEEE Computer Society Conference (2010)
3. Dorigo, M., Maniezzo, V., Coloni, A.: The ant system: optimization by a colony of cooperating agents. IEEE Trans. Syst. Man Cybern. B **26**(1), 29–41 (1996)
4. Gaxiola, F., Melin, P., López, M.: Modular neural networks for person recognition using segmentation and the iris biometric measurement with image pre-processing. Proceeding of InternationalJoint Conference on Neural Network, pp. 1–7 (2010)
5. Holliday, D., Resnick, R., Walker, J.: Fundamentals of Physics. Wiley, New York (1993)
6. Kang, S.C., Hong, S.: A speckle reduction filter using wavelet-based methods for medical imaging application. International Conference on Digital Signal Processing (DSP), pp. 1169–1172 (2002)
7. Kennedy, J., Eberhart, R.: Particle swarm optimization. Proc. IEEE Int. Conf. Neural Netw. **4**, 1942–1948 (1995)
8. Kirkpatrick, S., Gelatto, C.D., Vecchi, M.P.: Optimization by simulated annealing. Science **220**, 671–680 (1983)
9. Lemos, M., Navaux, A., Nehme, D., Olvier, P.: Cardiac structure recognition ultrasound images. Conference focused on Speech and Image Processing, Multimedia Communications and Services, pp. 463–466 (2007)
10. Martinez, G., Melin, P., Castillo, O.: Optimization of modular neural networks using hierarchical genetic algorithms applied to speech recognition. Proceeding of International Joint Conference on Neural Network, Montreal, Canada, 31 July–4 Aug 2005, pp. 1400–1406
11. Pandey, B., Garg, N.: Swarm optimized modular neural network based diagnostic system for breast cancer diagnosis. Int. J. Soft Comput. Artif. Intell. Appl. (IJSCAI) **2**(4) (2013)

12. Rashedi, E., Nezamabadi, H., Saryazdi, S.: GSA: a gravitational search algorithm. Inf. Sci. **179** (13), 2232–2248 (2009)
13. Tang, K.S., Man, K.F., Kwong, S., He, Q.: Genetic algorithms and their applications. IEEE Signal Process. Mag. **13**(6), 22–37 (1996)
14. Tsai, H., Lin, Y.: Modular Neural Network Programming with Genetic Optimization. Elsevier, Amsterdam, pp. 11032–11039
15. Valdez, F., Melin, P., Parra, H.: Parallel genetic algorithms for optimization of modular neural networks in pattern recognition. Proceeding of International Joint Conference on Neural Networks, San Jose, California, 31 July–5 Aug 2011, pp. 314–319

Ensemble Neural Network Optimization Using the Particle Swarm Algorithm with Type-1 and Type-2 Fuzzy Integration for Time Series Prediction

Martha Pulido and Patricia Melin

Abstract This paper shows the optimization of ensemble neural networks using the Particle Swarm algorithm for time series prediction with Type-1 and Type-2 Fuzzy Integration. The time series that is being considered in this paper is the Dow Jones Time Series. Simulation results show that the ensemble approach produces good prediction of the Dow Jones time series.

Keywords Ensemble neural networks · Particle swarm · Optimization · Time series prediction

1 Introduction

Time Series is called a set of measurements of some phenomenon or experiment recorded sequentially in time. The first step in analyzing a time series is to plot it, this allows: to identify the trend, seasonal, irregular variations. A classic model for a time series can be expressed as a sum or product of three components: trend, seasonality and random error term.

Time series predictions are very important because based on them we can analyze past events to know the possible behavior of futures events and thus we can take preventive or corrective decisions to help avoid unwanted circumstances [1, 2].

2 Optimization

In mathematics, computer science, or management science, mathematical optimization (alternatively, optimization or mathematical programming) is the selection of a best element (with regard to some criteria) from some set of available alternatives.

M. Pulido · P. Melin (✉)
Tijuana Institute of Technology, Tijuana, México
e-mail: pmelin@tectijuana.mx

© Springer International Publishing Switzerland 2015
O. Castillo and P. Melin (eds.), *Fuzzy Logic Augmentation of Nature-Inspired
Optimization Metaheuristics*, Studies in Computational Intelligence 574,
DOI 10.1007/978-3-319-10960-2_9

139

In the simplest case, an optimization problem consists of maximizing or minimizing a real function by systematically choosing input values from within an allowed set and computing the value of the function. The generalization of optimization theory and techniques to other formulations comprises a large area of applied mathematics. More generally, optimization includes finding "best available" values of some objective function given a defined domain (or a set of constraints), including a variety of different types of objective functions and different types of domains.

An optimization problem can be represented in the following way:

Given   a function f : A → R from some set A to the real numbers
Sought an element x_0 in A such that $f(x_0) \leq f(x)$ for all x in A ("minimization") or
          such that $f(x_0) \geq f(x)$ for all x in A ("maximization")

Such a formulation is called an optimization problem or a mathematical programming problem (a term not directly related to computer programming, but still in use for example in linear programming). Many real-world and theoretical problems may be modeled in this general framework. Problems formulated using this technique in the fields of physics and computer vision may refer to the technique as energy minimization, speaking of the value of the function f as representing the energy of the system being modeled.

Typically, A is some subset of the Euclidean space R^n, often specified by a set of *constraints*, equalities or inequalities that the members of A have to satisfy. The domain A of f is called the *search space* or the *choice set*, while the elements of A are called *candidate solutions* or *feasible solutions*.

The function f is called, variously, an objective function, a loss function or cost function (minimization), indirect utility function (minimization) [3], a utility function (maximization), a fitness function (maximization), or, in certain fields, an energy function, or energy functional. A feasible solution that minimizes (or maximizes, if that is the goal) the objective function is called an *optimal solution*.

By convention, the standard form of an optimization problem is stated in terms of minimization. Generally, unless both the objective function and the feasible region are convex in a minimization problem, there may be several local minima, where a *local minimum* x^* is defined as a point for which there exists some $\delta > 0$ so that for all x such that

$$\|x - x^*\| \leq \delta;$$

the expression

$$f(x^*) \leq f(x)$$

holds; that is to say, on some region around x^* all of the function values are greater than or equal to the value at that point. Local máxima are defined similarly [4–11].

3 Particle Swarm Optimization

The Particle Swarm Optimization algorithm maintains a swarm of particles, where each particle represents a potential solution. In analogy with evolutionary computation paradigms, a swarm is a population, while a particle is similar to an individual. In simple terms, the particles are "flown" through a multidimensional search space where the position of each particle is adjusted according to its own experience and that of their neighbors. Let $x_i(t)$ denote the position of particle i in the search space at time step t unless otherwise selected, t denotes discrete time steps. The position of the particle is changed by adding a velocity, $v_i(t)$. to the current position i,e.

$$x_i(t + 1) = x_i(t) + v_i(t + 1) \tag{1}$$

with

$$x_i(0) \sim U(X_{\min}, X_{\max}).$$

It is the velocity vector the one that drivers of the optimization process, and reflects both the experimental knowledge of the particles and the information exchanged in the vicinity of particles. The experimental knowledge of a particle which is generally known as the cognitive component, which is proportional to the distance of the particle from its own best position (hereinafter, the personal best position particles) that are from the first step. Socially exchanged information is known as the social component of the velocity equation.

For the best PSO, the particle velocity is calculated as:

$$v_{ij}(t + 1) = v_{ij}(\tau) + c_1 r_1 [y_{ij}(t) - x_{ij}(t)], + c_2 r_2(t)[\hat{y}_j(t) - x_{ij}(t)] \tag{2}$$

where $v_{ij}(t)$ is the velocity of the particle i in dimension j at time step t, $c_1 y c_2$ are positive acceleration constants used to scale the contribution of cognitive and social skills, respectively, y $r_{1j}(t)$, y $r_{2j}(t) \sim U(0, 1)$ are random values in the range [0,1].

The best personal position in the next time step $t + 1$ is calculated as:

$$y_i(t + 1) = \begin{cases} y_i(t) & \text{if } f(x_i(x_i(t + 1)) \geq f y_i(t)) \\ x_i(t + 1) & \text{if } f(x_i(x_i(t + 1)) > f y_i(t)) \end{cases}. \tag{3}$$

where $f : \mathbb{R}^{nx} \to \mathbb{R}$ is the fitness function, as with EAs, measuring fitness with the function will help find the optimal solution, for example the objective function quantifies the performance, or the quality of a particle (or solution).

The overall best position, $\hat{y}(t)$ at time step t, s defined as:

$$\hat{y}(t) \in \{y_o(t), \ldots, y_{ns}(t) f(y(t)\} \, f(y(t)) = \min\{y_o(t), \ldots, y_{ns}(t) f(y(t),\} \tag{4}$$

where n_s is the total number of particles in the swarm. Importantly, the above equation defining and establishing $\hat{y}$ the best position is uncovered by either of the particles so far as this is usually calculated from the best position best personal [12–15],

The overall best position may be selected from the actual swarm particles, in which case:

$$\hat{y}(t) = \min\{f(x_o(t)), \ldots f(x_{ns}(t)),\} \qquad (5)$$

4 Fuzzy Systems as Methods of Integration

Fuzzy logic was proposed for the first time in the mid-sixties at the University of California Berkeley by the brilliant engineer Lofty A. Zadeh., who proposed what it's called the principle of incompatibility: "As the complexity of system increases, our ability to be precise instructions and build on their behavior decreases to the threshold beyond which the accuracy and meaning are mutually exclusive characteristics." Then introduced the concept of a fuzzy set, under which lies the idea that the elements on which to build human thinking are not numbers but linguistic labels. Fuzzy logic can represent the common knowledge as a form of language that is mostly qualitative and not necessarily a quantity in a mathematical language [16].

Type-1 Fuzzy system theory was first introduced by Zadeh [17] in 1965, and has been applied in many areas such as control, data mining, time series prediction, etc.

The basic structure of a fuzzy inference system consists of three conceptual components: a rule base, which contains a selection of fuzzy rules, a database (or dictionary) which defines the membership functions used in the rules, and reasoning mechanism, which performs the inference procedure (usually fuzzy reasoning) [18].

Type-2 Fuzzy systems were proposed to overcome the limitations of a type-1 FLS, the concept of type-1 fuzzy sets was extended into type-2 fuzzy sets by Zadeh in 1975. These were designed to mathematically represent the vagueness and uncertainty of linguistic problems; thereby obtaining formal tools to work with intrinsic imprecision in different type of problems; it is considered a generalization of the classic set theory. Type-2 fuzzy sets are used for modeling uncertainty and imprecision in a better way [19, 20, 21, 22, 23, 24, 25, 26, 16].

5 Problem Statement and Proposed Method

The objective of this work is to develop a model that is based on integrating the responses of an ensemble neural network using type-2 fuzzy systems and optimization. Figure 1 represents the general architecture of the proposed method, where historical data, analyzing data, create the ensemble neural network and integrate responses of the ensemble neural network with type-2 fuzzy system integration and we obtaining the output are shown. The information can be historical data, these can be images, time series, etc., in this case we show the application to time series prediction of the Dow Jones where we obtain good results with this series.

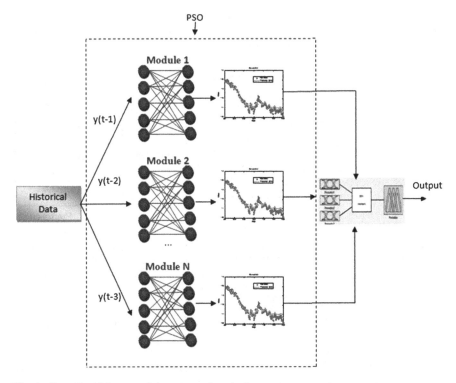

Fig. 1 General architecture of the proposed method

Figure 2 shows a type-2 fuzzy system consisting of with 5 inputs depending on the number of modules of the neural network ensemble and one output. Each input and output linguistic variable of the fuzzy system uses 2 Gaussian membership functions. The performance of the type-2 fuzzy integrators is analyzed under different levels of uncertainty to find out the best design of the membership functions and consist of 32 rules. For the type-2 fuzzy integrator using 2 membership functions which are called low prediction and high prediction for each of the inputs and output of the fuzzy system. The memberships functions are of Gaussian type, we consider 3 sizes for the footprint uncertainty 0.3, 0.4 and 0.5 to obtain a better prediction of our time series.

Figure 3 shows the possible rules of a type-2 fuzzy system.

Figure 4 represents the Particle Structure to optimize the ensemble neural network, where the parameters that are optimized is the number de modules, number of layers, number of neurons.

Historical data of the Dow Jones time series was used for the ensemble neural network trainings, where each module was fed with the same information, unlike modular networks, where each module is fed with different data, which leads to architectures that are not uniform.

Fig. 2 Type 2 fuzzy system
for DowJones time series

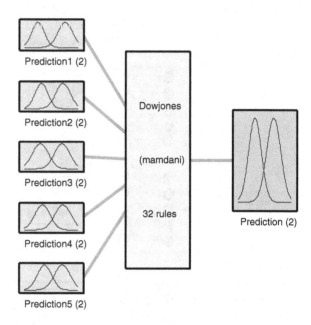

System Dowjones: 5 inputs, 1 outputs, 32 rules

The Dow Jones (DJ) is a U.S. company that publishes financial information. Founded in 1882 by three reporters: Charles Henry Dow, Edward David Jones and Charles Milford Bergstresser.

In the same year it began publishing a financial newsletter called "The Customer's Afternoon Letter which "would be the precursor of the famous financial newspaper The Wall Street Journal first published on July 8, 1889.

The newsletter showed publicly share prices and the financial accounts of companies, information that until then had only the people close to the companies.

To better represent the movements of the stock market at the time, the Dow Jones designed a barometer of economic activity meter with twelve companies creating the Dow Jones stock index.

Like the New York Times and the Washington Post newspapers, the company is open to the market but is controlled the by the private sector. So far, the company is controlled by the Bancroft family, which controls 64 % of the shares entitled to vote [12].

Data of the Dow Jones time series: We are using 800 points that correspond to a period from 08/12/2008 to 09/09/2011 (as shown in Fig. 4). We used 70 % of the data for the ensemble neural network trainings and 30 % to test the network [13].

1. If (Prrediction1 is Pred1Low) and (Prediction2 is Pron2Low) and (Prediction3 is Pred3Low) and (Prediction4 is Pred4Low) and (Prediction5is Pred5Low) then (Prediction is Low)
2. If (Prrediction1 is Pred1High) and (Prediction2 is Pron2High) and (Prediction3 is Pred3High) and (Prediction4 is Pred4High) and (Prediction5is Pred5High) then (Prediction is High)
3. If (Prrediction1 is Pred1Low) and (Prediction2 is Pron2Low) and (Prediction3 is Pred3Low) and (Prediction4 is Pred4Low) and (Prediction5is Pred5Low) then (Prediction is Low)
4. If (Prrediction1 is Pred1High) and (Prediction2 is Pron2High) and (Prediction3 is Pred3High) and (Prediction4 is Pred4High) and (Prediction5is Pred5High) then (Prediction is High)
5. If (Prrediction1 is Pred1Low) and (Prediction2 is Pron2Low) and (Prediction3 is Pred3Low) and (Prediction4 is Pred4High) and (Prediction5is Pred5High) then (Prediction is Low)
6. If (Prrediction1 is Pred1High) and (Prediction2 is Pron2High) and (Prediction3 is Pred3High) and (Prediction4 is Pred4Low) and (Prediction5is Pred5Low) then (Prediction is High)
7. If (Prrediction1 is Pred1Low) and (Prediction2 is Pron2Low) and (Prediction3 is Pred3High) and (Prediction4 is Pred4High) and (Prediction5is Pred5High) then (Prediction is High)
8. If (Prrediction1 is Pred1High) and (Prediction2 is Pron2High) and (Prediction3 is Pred3Low) and (Prediction4 is Pred4Low) and (Prediction5is Pred5Low) then (Prediction is Low)
9. If (Prrediction1 is Pred1Low) and (Prediction2 is Pron2High and (Prediction3 is Pred3High) and (Prediction4 is Pred4High) and (Prediction5is Pred5High) then (Prediction is High)
10. If (Prrediction1 is Pred1High) and (Prediction2 is Pron2Low) and (Prediction3 is Pred3Low) and (Prediction4 is Pred4Low) and (Prediction5is Pred5Low) then (Prediction is Low)
11. If (Prrediction1 is PredLow) and (Prediction2 is Pron2High) and (Prediction3 is Pred3Low) and (Prediction4 is Pred4High) and (Prediction5is Pred5Low) then (Prediction is Low)
12. If (Prrediction1 is Pred1High) and (Prediction2 is Pron2Low) and (Prediction3 is Pred3High) and (Prediction4 is Pred4Low) and (Prediction5is Pred5High) then (Prediction is High)
13. If (Prrediction1 is PredLow) and (Prediction2 is Pron2High) and (Prediction3 is Pred3Low) and (Prediction4 is PredLow) and (Prediction5is Pred5Low) then (Prediction is Low)
14. If (Prrediction1 is Pred1High) and (Prediction2 is Pron2Low) and (Prediction3 is Pred3High) and (Prediction4 is Pred4Low) and (Prediction5is Pred5High) then (Prediction is High)
15. If (Prrediction1 is PredLow) and (Prediction2 is Pron2High) and (Prediction3 is Pred3Low) and (Prediction4 is PredLow) and (Prediction5is Pred5Low) then (Prediction is Low)
16. If (Prrediction1 is Pred1High) and (Prediction2 is Pron2Low) and (Prediction3 is Pred3High) and (Prediction4 is Pred4Low) and (Prediction5is Pred5High) then (Prediction is High)
17. If (Prrediction1 is PredLow) and (Prediction2 is Pron2Low) and (Prediction3 is Pred3High) and (Prediction4 is PredLow) and (Prediction5is Pred5Low) then (Prediction is Low)
18. If (Prrediction1 is Pred1High) and (Prediction2 is Pron2High) and (Prediction3 is Pred3Lowh) and (Prediction4 is Pred4Low) and (Prediction5is Pred5High) then (Prediction is High)
19. If (Prrediction1 is PredLow) and (Prediction2 is Pron2Low) and (Prediction3 is Pred3High) and (Prediction4 is Pred4High) and (Prediction5is Pred5High) then (Prediction is Low)
20. If (Prrediction1 is Pred1High) and (Prediction2 is Pron2High) and (Prediction3 is Pred3Lowh) and (Prediction4 is Pred4Low) and (Prediction5is Pred5High) then (Prediction is High)
21. If (Prrediction1 is Pred1Low) and (Prediction2 is Pron2High) and (Prediction3 is Pred3High) and (Prediction4 is Pred4Low) and (Prediction5is Pred5Low) then (Prediction is Low)
22. If (Prrediction1 is Pred1High) and (Prediction2 is Pron2Low) and (Prediction3 is Pred3Low) and (Prediction4 is Pred4High) and (Prediction5is Pred5High) then (Prediction is High)
23. If (Prrediction1 is Pred1Low) and (Prediction2 is Pron2Low) and (Prediction3 is Pred3High) and (Prediction4 is Pred4High) and (Prediction5is Pred5Low) then (Prediction is Low)
24. If (Prrediction1 is Pred1High) and (Prediction2 is Pron2High) and (Prediction3 is Pred3Low) and (Prediction4 is Pred4Low) and (Prediction5is Pred5High) then (Prediction is High)
25. If (Prrediction1 is Pred1Low) and (Prediction2 is Pron2High) and (Prediction3 is Pred3High) and (Prediction4 is Pred4Low) and (Prediction5is Pred5High) then (Prediction is High)
26. If (Prrediction1 is Pred1Low) and (Prediction2 is Pron2High) and (Prediction3 is Pred3Low) and (Prediction4 is Pred4High) and (Prediction5is Pred5Low) then (Prediction is Low)
27. If (Prrediction1 is Pred1Low) and (Prediction2 is Pron2High) and (Prediction3 is Pred3Low) and (Prediction4 is Pred4High) and (Prediction5is Pred5Low) then (Prediction is Low)
28. If (Prrediction1 is Pred1Low) and (Prediction2 is Pron2High) and (Prediction3 is Pred3Low) and (Prediction4 is Pred4High) and (Prediction5is Pred5Low) then (Prediction is Low)
29. If (Prrediction1 is Pred1Low) and (Prediction2 is Pron2High) and (Prediction3 is Pred3High) and (Prediction4 is Pred4High) and (Prediction5is Pred5Low) then (Prediction is High)
30. If (Prrediction1 is Pred1High) and (Prediction2 is Pron2Low) and (Prediction3 is Pred3Low) and (Prediction4 is Pred4Low) and (Prediction5is Pred5High) then (Prediction is Low)
31. If (Prrediction1 is Pred1Low) and (Prediction2 is Pron2High) and (Prediction3 is Pred3Low) and (Prediction4 is Pred4High) and (Prediction5is Pred5High) then (Prediction is High)
302 If (Prrediction1 is Pred1High) and (Prediction2 is Pron2Low) and (Prediction3 is Pred3High) and (Prediction4 is Pred4Low) and (Prediction5is Pred5Low) then (Prediction is Low)

Fig. 3 Rules of type-2 fuzzy inference system of the Dow Jones time series

Number of Modules	Number of Layers 1	Neurons 1		Neurons ... n

Fig. 4 Particle structure to optimize the ensemble neural network

6 Simulation Results

In this section we present the simulation results obtained with the integration of ensemble neural networks with type-1 and type-2 fuzzy integration and optimization with the Particle Swarm for the Dow Jones time series.

Table 1 shows the particle swarm results (as shown in Fig. 5) where the prediction error is of 0.0037691 (Table 2).

Table 1 Particle Swarm results for the ensemble neural network

No.	Iterations	Particles	Number of modules	Number of layers	Number of neurons	Duration	Prediction error
1	100	100	3	2	19,9 19,21 8,11	02:22:18	0.0056306
2	100	100	4	2	18,25 16,16 21,23 6,24	02:40:22	0.0058939
3	100	100	2	1	3 5	03:11:05	0.005457
4	100	100	2	2	8,14 5,4	01:32:11	0.0037691
5	100	100	3	3	3,12,13 10,27,26 8,17,11	01:20:27	0.0046956
6	100	100	3	2	23,13 16,17 11,11	02:06:15	0.0044795
7	100	100	3	3	17,11,22 21,24,21 8,21,11	01:09:03	0.0050147
8	100	100	2	2	20,26 6,4	01:26:20	0.0041311
g	100	100	4	3	8,21,21 21,22,16 15,16,9	01:37:07	0.0048197
10	100	100	3	2	23,7 4,20 11,21	02:00:41	0.0045441

Fig. 5 Dow Jones time series 1

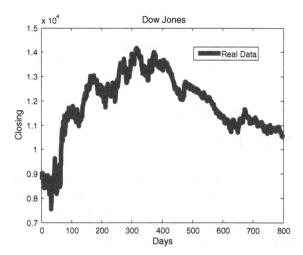

Table 2 Results of type-1 fuzzy integration of DJ

Experiment	Integration type-1
Experiment 1	0.3972
Experiment 2	0.0538
Experiment 3	0.2642
Experiment 4	0.0608
Experiment 5	0.0866
Experiment 6	0.1184
Experiment 7	0.24589
Experiment 8	0.1541
Experiment 9	0.1186
Experiment 10	0.1152

Table 3 Results of type-2 fuzzy integration of DJ

Experiment	Prediction error 0.3 uncertainty	prediction error 0.4 uncertainty	Prediction error 0.5 uncertainty
Experiment 1	0.4284	0.3961	0.3522
Experiment 2	0.3257	0.3152	0.3089
Experiment 3	0.3580	0.2121	0.2126
Experiment 4	0.0528	0.0421	0.0445
Experiment 5	0.0456	0.0432	0.029
Experiment 6	0.0352	0.0348	0.0258
Experiment 7	0.0397	0.0356	0.0347
Experiment 8	0.0080	0.0653	0.0621
Experiment 9	0.0222	0.02145	0.0205
Experiment 10	0.0437	0.0428	0.0418

Fuzzy integration is also performed by implementing a type-2 fuzzy system in which the results were as follows: for the best evolution with a degree of uncertainty of 0.3 a forecast error of 0.0222 was obtained, and with a degree of uncertainty of 0.4 the error is of: 0.02145 and with a degree uncertainty 0.5 the error of 0.0205, as shown in Table 3.

7 Conclusions

In this paper we considered the PSO algorithm to optimize the architecture of the ensemble neural network to predict the time series of the Dowjones, where good results were obtained, and we can say that this algorithm in good in speed when compared with other optimization techniques PSO is an effective and efficient metahuristic to find the solution of problems. In conclusion the use of ensemble neural networks with type-2 fuzzy integration could be a good choice in predicting complex time series. Future works would be was optimize with particles swarm the type-1 and type-2 fuzzy systems to obtain a better prediction error. Type-1 and Type-2 fuzzy system could optimized in terms of the parameter of membership functions, membership type and number of rules.

Acknowledgment We would like to express our gratitude to the CONACYT, Tijuana Institute of Technology for the facilities and resources granted for the development of this research.

References

1. Brockwell, P.D., Richard, A.D.: Introduction to Time Series and Forecasting, pp. 1–219. Springer, New York (2002)
2. Krogh, A., Vedelsby, J.: Neural network ensembles, cross validation, and active learning. In: Tesauro, G., Touretzky, D., Leen, T. (eds.), Advances in Neural Information Processing Systems, vol. 7, pp. 231–238. MIT Press, Cambridge (1995)
3. Castillo, O., Melin, P.: Hybrid intelligent systems for time series prediction using neural networks, fuzzy logic, and fractal theory neural networks. IEEE Trans. **13**(6), 1395–1408 (2002)
4. Golberg, D. (ed.): Genetic Algorithms in Search, Optimization and Machine Learning. Addison Wesley (1989)
5. Plummer, E.A. Time Series Forecasting with Feed-Forward Neural Networks: Guidelines and Limitations. University of Wyoming 2000
6. Maguire, L.P., Roche, B., McGinnity, T.M., McDaid, L.J.: Predicting a chaotic time series using a fuzzy neural network. **12**(1–4), 125–136 (1998)
7. Multaba, I.M., Hussain, M.A. Application of Neural Networks and Other Learning. Technologies in Process Engineering. Imperial Collage Press (2001)
8. Sollich P., Krogh, A., Learning with ensembles: how over-fitting can be useful. In: Touretzky, D.S., Mozer, M.C., Hasselmo, M.E. (eds.) Advances in Neural Information Processing Systems 8, Denver, CO, pp. 190–196, MIT Press, Cambridge (1996)
9. Yadav, R.N., Kalra, P.K., John, J.: Time series prediction with single multiplicative neuron model. Soft Comput. Time Ser. Prediction Appl. Soft Comput. **7**(4), 1157–1163 (2007)
10. Yao, X., Liu, Y.: Making use of population information in evolutionary artificial neural networks. IEEE Trans. Syst. Man and Cybern. Part B: Cybern. **28**(3), 417–425 (1998)
11. Zhou, Z.-H., Jiang, Y., Yang, Y.-B., Chen, S.-F.: Lung cancer cell identification based on artificial neural network ensembles. Artif. Intell. Med. **24**(1), 25–36 (2002)
12. Dow Jones Company. http://www.dowjones.com. New York, 10 Jan 2010
13. Dow Jones Indexes. http://www.djindexes.com 5 Sept 2010
14. Karnik, N., Mendel, M.: Applications of type-2 fuzzy logic systems to forecasting of time-series. Inf. Sci. **120**(1–4), 89–111 (1999)

15. Zhao, L. Yang, Y.: PSO-based single multiplicative neuron model for time series prediction. Expert Syst. Appl. **36**(2), 2805-2812 (2009)
16. Reeves, R.C., Row E.J.: Genetic Algorithms: Principles and Perspectives, A Guide to GA Theory, pp. 2–17, Kluwer Academic Publishers, New York (2003)
17. Zadeh, L.A.: Fuzzy sets and applications: selected papers. In: R.R. Yager et al. (eds.). Wiley, New York (1987)
18. Pulido, M., Mancilla, A., Melin, P.: An Ensemble Neural Network Architecture with Fuzzy Response Integration for Complex Time Series Prediction Evolutionary Design of Intelligent Systems in Modeling, Simulation and Control. pp. 85–110, Springer, Berlin (2009)
19. Sharkey, A.: One combining Artificial of Neural Nets. Department of Computer Science University of Sheffield, UK (1996)
20. Sharkey, A.A.: Combining Artificial Neural Nets: Ensemble and Modular Multi-Net Systems. Springer, London (1999)
21. Shimshoni, Y.N.: Initiator Classification of seismic signal by integrating ensemble of neural networks. IEEE Trans. Sig. Process. **461**(5), 1194–1201 (1998)
22. Castillo, O., Melin, P.: Type-2 Fuzzy Logic: Theory and Applications Neural Networks, pp. 30–43, Springer (2008)
23. Castillo, O., Melin, P.: Simulation and forecasting complex economic time series using neural networks and fuzzy logic, Proceedings of the International Neural Networks Conference, vol. 3, pp. 1805–1810. San Diego (2001)
24. Castillo, O., Melin P.: Simulation and forecasting complex financial time series using neural networks and fuzzy logic, Proceedings of the IEEE International Conference on Systems, Man and Cybernetics, vol. 4, pp. 2664–2669. Tucson, AZ (2001)
25. Jang, J.S.R, Sun, C.T. Mizutani, E.: Neuro-Fuzzy and Sof Computing. Prentice Hall, New Jersey (1996)
26. Castillo, P.O., Gonzalez, S., Cota, J., Trujillo, W., Osuna, P.: Design of Modular Neural Networks with Fuzzy Integration Applied to Time Series Prediction. vol. 41, pp. 265–273, Springer, Berlin (2007)

Clustering Bin Packing Instances for Generating a Minimal Set of Heuristics by Using Grammatical Evolution

Marco Aurelio Sotelo-Figueroa, Héctor José Puga Soberanes,
Juan Martín Carpio, Héctor J. Fraire Huacuja, Laura Cruz Reyes
and Jorge Alberto Soria Alcaraz

Abstract Grammatical Evolution has been used to evolve heuristics for the Bin Packing Problem. It has been shown that the use of Grammatical Evolution can generate an heuristic for either one instances or a full instance set for this problem. In many papers the selection of instances for heuristics generation has been done randomly. The present work proposes a methodology to cluster bin packing instances and choose the instances to generate an heuristic for each cluster. The number of heuristics generated is based on the number of clusters. There were used only one instance by cluster. The results obtained were compared through non-parametric tests against the best known heuristics.

Keywords Grammatical Evolution · Bin Packing Problem · Heuristics

1 Introduction

The *Heuristics* [1, 2] are defined as "a type of strategy that dramatically limits the search for solutions" meanwhile the *Metaheuristics* [3] are defined as "a master strategy that guides and modifies other heuristics to obtain solutions generally better

M.A. Sotelo-Figueroa (✉) · H.J. Puga Soberanes · J.M. Carpio · J.A. Soria Alcaraz
Instituto Tecnológico de León, León, GTO, Mexico
e-mail: marco.sotelo@itleon.edu.mx

H.J. Puga Soberanes
e-mail: pugahector@yahoo.com

J.M. Carpio
e-mail: jmcarpio61@hotmail.com

J.A. Soria Alcaraz
e-mail: soajorgea@gmail.com

H.J. Fraire Huacuja · L.C. Reyes
Instituto Tecnológico de Ciudad Madero, Tamaulipas, Mexico
e-mail: hfraire@prodigy.net.mx

L.C. Reyes
e-mail: lcruzreyes@prodigy.net.mx

© Springer International Publishing Switzerland 2015
O. Castillo and P. Melin (eds.), *Fuzzy Logic Augmentation of Nature-Inspired Optimization Metaheuristics*, Studies in Computational Intelligence 574,
DOI 10.1007/978-3-319-10960-2_10

that the ones obtained with a local search optimization". One important charac teristic of heuristics is that they can obtain a result for an instance problem in polynomial time [4], although heuristics are developed for a specific instance problem. Metaheuristics however can work over several instances of a given problem or various problems, but it is necessary to adapt the metaheuristics to work with each problem.

It has been shown that metaheuristics, like Genetic Programming [5], can generate an heuristic that can be applied to an instance problem [6]. However there are others metaheuristics that work like the Genetic Programming's paradigm [7] such as *Grammatical Differential Evolution* [8], *Grammatical Swarm* [9] and others [10–12] and is possible to generate an heuristics with them.

In [6, 12–16] was generated a heuristics for a problem, with each instance set they tried to generate an heuristic. To do it was chosen one instance to train the metaheuristic, but the instance was chosen randomly among the instance set.

In the present paper we will work with the Bin Packing Problem because it is a problem that has been amply studied and that has generated heuristics [17–20] and metaheuristics [21–23] to try to get betters results.

In [24, 25] was proposed a metric to cluster the Bin Packing Instances, such metric was based on generalize the Bin Packing Problem and the item's weights of each instance. It showed that was possible to cluster the instances into three cases, and was showed how to determinate which instances from each cluster is hardest for common algorithms.

The heuristics generated by the clustering the Bin Packing Instances using Grammatical Evolution were compared to those generated by [26] using the non-parametric Friedman Test [27], the quality of each heuristic was based on the free space percentage obtained through the fitness function.

The aim of the investigation is to prove that by mean of the Schwerin's metric is possible to cluster bin packing instances, for generating cluster representative heuristics by using metaheuristic algorithms; this removes the redundant generation of heuristics. This phenomena have been observed in heuristic generation methods that lack of the clustering process even using the same metaheuristic algorithm.

2 Bin Packing Problem

The classical one dimensional bin packing problem (BPP) [28] consists of a set of pieces, which must be packed into as few bins as possible. Each piece j has a weight w_j, and each bin has capacity c. The objective is to minimise the number of bins used, where each piece is assigned to one bin only, and the weight of the pieces in each bin does not exceed c. This NP-complete decision problem naturally gives rise to the associated NP-hard optimization problem.

A mathematical definition of *Bin Packing Problem* [28, 29] is:
Minimize:

$$z = \sum_{i=1}^{n} y_i \tag{1}$$

Subject to:

$$\sum_{j=1}^{n} w_j x_{ij} \leq c y_i \quad i \in N = \{1, \ldots, n\} \tag{2}$$

$$\sum_{i=1}^{n} x_{ij} = 1 \quad j \in N \tag{3}$$

$$y_i \in \{0, 1\} \quad i \in N \tag{4}$$

$$x_{ij} \in \{0, 1\} \quad i \in N, \quad j \in N \tag{5}$$

where:

w_j weight of item j
y_j binary variable that shows if the bin i have items
x_{ij} indicates whether the item j is in the bin i
n number of available bins (also the number of items n)
c capacity of each bin

2.1 Instances

Schoenfield proposed the *Hard28* [30]. It is considered like the most difficult instances which could not be solved neither by the pure cutting plane algorithm from [31] nor by many reduction methods.

Beasley [32] proposed a collection of test data sets, known as *OR-Library* and maintained by the Beasley University, which were studied by Falkenauer [22]. This collection contains a variety of test data sets for a variety of Operational Research problems, as the BPP in several dimensions. For the one dimensional BPP the collection contains eight data sets, that can be classified in two classes:

- *Unifor* The data sets from binpack1 to binpack4 consist of items of sizes uniformly distributed in $(200, 100)$ to be packed into bins of size 150. The number of bins in the current known solution was found by [22].
- *Triplets* The data sets from binpack5 to binpack8 consist of items from $(24, 50)$ to be packed into bins of size 100. The number of bins can be obtained dividing the size of the data set by three.

Scholl [33] proposed another collection of data sets and was report 1184 of the problems have been solved to optimality. Alvim [34] reported the optimal solutions for the remaining 26 problems. The collection contains three data sets:

Set 1 It has 720 instances with items drawn from a uniform distribution on three intervals $[1, 100]$, $[20, 100]$, and $[30, 100]$. The bin capacity is C = 100, 120, and 150 and n = 50, 100, 200, and 500.

Set 2 It has 480 instances with C = 1000 and n = 50, 100, 200, and 500. Each bin has an average of 3–9 items.

Set 3 It has 10 instances with C = 100,000, n = 200, and items are drawn from a uniform distribution on $[20000, 35000]$. Set 3 is considered the most difficult of the three sets.

2.2 Fitness Measure

In [35] was propose an objective function which puts a premium on bins that are filled completely or nearly so. Importantly, the fitness function is designed to avoid the problem of plateaus in the search space, that occur when the fitness function does not discriminate between heuristics that use the same number of bins.

$$Fitness = 1 - \left(\frac{\sum_{i=1}^{n} \left(\frac{\sum_{j=1}^{m} w_j x_{ij}}{c} \right)^2}{n} \right) \tag{6}$$

where:

n	number of bins
m	number of item
w_j	weight of item j-th
x_{ij}	$x_{ij} = \begin{cases} 1 & \text{if the item } j \text{ is in the bin} \\ 0 & \text{otherwise} \end{cases}$
c	bin capacity

2.3 Heuristics

To solve the Bin Packing Problem are many heuristics like the Best-Fit [19]. It is a constructive heuristic which packs a set of pieces one at a time, in the order that they are presented. The heuristic sorts the bins and iterates through the open bins, and the current piece is placed in the first bin into which it fits. This heuristic is for the online bin packing problem, because it packs pieces one at a time, and a piece cannot be moved once it has been assigned to a bin. The Best-Fit algorithm can be seen in the algorithm 1.

Algorithm 1 Best-Fit Algorithm
___
Require: L input list of items to be packed.
 1: Initialize s with a empty bin.
 2: **for all** items p in L **do**
 3:     Sort the bins, ordering by free space.
 4:     **for all** bins b in s **do**
 5:         **if** $Capacity(b) \leq FreeSpace(b) + Size(p)$ **then**
 6:             Put the item p in the bin b.
 7:         **end if**
 8:     **end for**
 9:     **if** p was not place in any bin **then**
10:         Add a new bin b to s and add the item p into the bin b.
11:     **end if**
12: **end for**

2.4 Clustering

The Bin Packing Problem doesn't have too much to parametrize apart the item's weight and the bin's capacity. Many researchers have investigated the range and weights and they have found that a smaller range makes the instance more difficult for common algorithms [24, 36–38].

Schwerin [24] showed that all the items inside the bin packing instance can be generated as follows:

$$w \in [v_L \chi, v_U \chi] \tag{7}$$

where:

w	item
v_L	lower bound
v_U	upper bound
χ	bin capacity

In [25, 37] was proposes to clustering the bin packing instances into three groups, as shown in the Table 1, such clusters were based on the Eq. (7).

Table 1 Bin packing instance clusters

Group	Condition
Triplets	$v_L = \frac{1}{4}$ and $v_U = \frac{1}{2}$
Hards	$\bar{w} \approx \frac{1}{3}$ or close to $\frac{1}{n}$ where $n \geq 3$
Regulars	the others

3 Grammatical Evolution

Grammatical Evolution (*GE*) [7] is a grammar-base form of Genetic Programming (*GP*) [39]. GE joins the principles from molecular biology, which are used by the GP, and the power of formal grammars. Unlike GP, the GE adopts a population of lineal genotypic integer strings, or binary strings, witch are transformed into functional phenotypic through a genotype-to-phenotype mapping process, this process is also know as *Indirect Representation* [40]. This transformation is governed through a Backus Naur Form grammar (*BNF*). Genotype strings are evolved with no knowledge of their phenotypic equivalent, only use the fitness measure.

3.1 Mapping Process

When approaching a problem using GE, initially a BNF grammar must be defined. This grammar specifies the syntax of desired phenotypic programs to be produced by GE. The development of a BNF grammar also affords the researcher the ability to incorporate domain biases or domain-specific functions.

A BNF grammar is made up of the tuple N, T, P, S; where N is the set of all non-terminal symbols, T is the set of terminals, P is the set of production rules that map $N \rightarrow T$, and S is the initial start symbol where $S \in N$. Where there are a number of production rules that can be applied to a non-terminal, a "|" (or) symbol separates the options.

Using the grammar as the GE input, the Eq. (8) is used to choose the next production based-on the non-terminal symbol.

$$Rule = c\%r \tag{8}$$

where c is the codon value and r is the number of production rules available for the current non-terminal.

An example of the mapping process employed by GE is shown in Fig. 1

4 Experiments

The proposed approach tries to evolve an heuristic for each cluster; to do this was joined the instances from all instances set and was applied the clustering process based on the Table 1. In [26] was proposed a Grammar to evolve heuristics and these heuristics obtained had the same performance of the Best-Fit heuristic (Fig. 2).

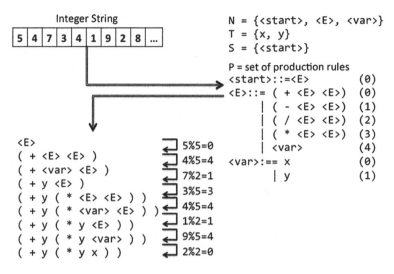

Fig. 1 An example a transformation from genotype to phenotype using a BNF Grammar. It begins with the start symbol, if the production rule from this symbol is only one rule, then the production rule gets instead of the start symbol, and the process begins to choose the productions rules base on the current genotype. It is taking each genotype and the non-terminal symbol from the *left* to realize the next production using the Eq. (8) until all the genotypes are mapped or there aren't more non-terminals in the phenotype

where:

S	size of the current piece
C	bin capacity
F	sum of the pieces already in the bin
Bin	sort the bins base on the bin number
Cont	sort the bins base on the bin contents
Asc	sort the elements in ascendant order
Des	sort the elements in descendant order

Fig. 2 Grammar proposal by [26] to obtain results like the obtained by the best-fist heuristic

$\langle inicio \rangle$ ♣ $\langle exprs \rangle$ ▷ ($\langle expr \rangle$) <=($\langle expr \rangle$)

$\langle exprs \rangle$ ♣ *Sort* (($\langle exprk \rangle$ᶜ $\langle order \rangle$) ♣λ

$\langle exprk \rangle$ ♣ Bin ♣ Content

$\langle order \rangle$ ♣ Asc ♣ Des

$\langle expr \rangle$ ♣ (($\langle expr \rangle$ $\langle op \rangle$ $\langle expr \rangle$) ♣ $\langle var \rangle$ ♣ *abs* (($\langle expr2 \rangle$))

$\langle expr2 \rangle$ ♣ (($\langle expr2 \rangle$ $\langle op \rangle$ $\langle expr2 \rangle$) ♣ $\langle var \rangle$

$\langle var \rangle$ ♣ F ♣ C ♣ S

$\langle op \rangle$ ♣ + ♣ * ♣ - ♣ /

Table 2 Initialization
parameters of each GE run

Parameter	Value
Population size	50
Differential weight	0.9
Crossover rate	1.0
Scheme	DE/best/1
Function calls	1,000

After an heuristic is generated for each cluster, it is applied to all cluster's instance. To compare the results between the different works was necessary to gather the instances into their original instances sets.

The fitness was calculated applying the objective function Eq. (6) to each instances and was added all the fitness into the instances set.

The metaheuristic's parameters used was taked from [26] in the who the author used a Covery Arrays to choose the right parameters, those parameters are shown in the Table 2.

For each set an heuristic was generated and it was chosen after performing 32 experiments and choose the median between them. The results were compared against the results obtained from the heuristics Best-Fit and the obtained by [26]. The comparison was implemented through the non-parametric test of Friedman, Aligned Friedman and Quade [27, 41]. Those non-parametric tests allow to discern if the raised hypothesis is true or false, and they use an post hoc analysis when the raised hypothesis is false.

5 Results

The Table 3 shows the heuristics generated by clustering the instances set. After was rejoined the intances into their original instance set was estimated the instance set's fitness, the results obtained are shown in the Table 4.

With the results obtained a non-parametric test was performed, in order to prove the hypothesis. Since the hypothesis proved to be false, the Table 5 shown the values and the p-values from each non-parametric test, was necessary applied the post hoc procedures to obtain the ranking of the heuristics, as shown in Table 6.

Table 3 Heuristics obtained
for each cluster

Cluster	Heuristic
Triplets	$(abs((F + S))) \leq (C)$
Hard	Sort (Bin, Asc). $((S + F)) \leq (abs(C))$
Regular	Sort (Cont, Des). $((F + abs(S))) \leq (C)$

Table 4 Results obtained

Instance	BestFit	GE	GE-Cluster
bin1data	44.561	44.561	44.561
bin2data	44.561	44.561	44.561
bin3data	1.3902	1.3902	1.3902
binpack1	2.4258	2.4258	2.4258
binpack2	2.2591	2.2591	2.2591
binpack3	2.0145	2.0145	2.0145
binpack4	1.8387	1.8387	1.8387
binpack5	0.0	0.0	0.0
binpack6	0.0	0.0	0.0
binpack7	0.0	0.0	0.0
binpack8	0.0	0.0	0.0
hard28	0.6555	0.6555	0.6555

Table 5 Non parametric test p-values

Non parametric test	Value	p-value
Friedman	0.0	1.0
Aligned Friedman	9.49732	0.00866
Quade	0.0	1.0

Table 6 Rankings of the algorithms

Algorithm	Ranking		
	Friedman	Aligned Friedman	Quade
BestFit	2.0	18.5	2.0
GE	2.0	18.5	2.0
GE-Cluster	2.0	18.5	2.0

6 Conclusions and Future Work

Based on the obtained results in Sect. 5, it is possible to conclude that it is possible to cluster the instances using Schwerin's metric and generate heuristics using GE.

The results obtained show that with only three heuristics applied to the Bin Packing Instances we can obtain the same performance that if we make one heuristic by each instance set.

Is necessary to search for a Grammar that gives better heuristics than the Best-Fit heuristic and analyze if others metaheuristics can be used with the Mapping Process.

The current investigation is based on the one dimensional bin packing problem but this methodology can be used to solve other problems, due to the generality of the approach, to this aim is necessary to look up for the right metric.

Acknowledgement Authors thanks the support received from Consejo Nacional de Ciencia y Tecnologia (CONACyT).The authors want to thank to *Instituto Tecnológico de León* (ITL) for the support to this research. Additionally they want to aknowledge the generous support from the *Mexican National Council for Science and Technology* (CONACyT) for this research project.

References

1. Feigenbaum, E.A., Feldman, J.: Computers and Thought. AAAI Press (1963)
2. Romanycia, M.H.J., Pelletier, F.J.: What is a heuristic? Comput. Intell. **1**(1), 47–58 (1985)
3. Glover, F.W.: Future paths for integer programming and links to artificial intelligence. Comput. Oper. Res. **13**, 533–549 (1986)
4. Garey, M.R., Johnson, D.S.: Computers and Intractability: A Guide to the Theory of NP-Completeness. W. H. Freeman & Co., New York, NY, USA (1979)
5. Koza, J.R.: Hierarchical genetic algorithms operating on populations of computer programs. In: IJCAI. pp. 768–774 (1989)
6. Burke, E.K., Hyde, M., Kendall, G.: Evolving bin packing heuristics with genetic programming. In: Runarsson, T., Beyer, H.G., Burke, E., Merelo-Guervós, J., Whitley, L., Yao, X. (eds.) Parallel Problem Solving from Nature—PPSN IX. Lecture Notes in Computer Science, vol. 4193, pp. 860–869. Springer, Berlin, Heidelberg (2006)
7. Ryan, C., Collins, J., Collins, J., O'Neill, M.: Grammatical evolution: Evolving programs for an arbitrary language. In: Proceedings of the First European Workshop on Genetic Programming, Lecture Notes in Computer Science 1391, pp. 83–95. Springer (1998)
8. M., O., A, B.: Grammatical differential evolution. In: International Conference on Artificial Intelligence (ICAI'06). CSEA Press, Las Vegas, Nevada (2006)
9. O'Neill, M., Brabazon, A.: Grammatical swarm: The generation of programs by social programming. Nat. Comput. **5**(4), 443–462 (2006)
10. Togelius, J., Nardi, R.D., Moraglio, A.: Geometric pso + gp = particle swarm programming. IEEE Congress on Evolutionary Computation, pp. 3594–3600 (2008)
11. Moraglio, A., Silva, S.: Geometric differential evolution on the space of genetic programs. In: Esparcia-Alcázar, A., Ekárt, A., Silva, S., Dignum, S., Uyar, A. (eds.) Genetic Programming. Lecture Notes in Computer Science, vol. 6021, pp. 171–183. Springer, Berlin / Heidelberg (2010)
12. Sotelo-Figueroa, M.A., Puga Soberanes, H.J., Martín Carpio, J., Fraire Huacuja, H.J., Reyes, C.L., Soria-Alcaraz, J.A.: Evolving bin packing heuristic using micro-differential evolution with indirect representation. In: Castillo, O., Melin, P., Kacprzyk, J. (eds.) Recent Advances on Hybrid Intelligent Systems, Studies in Computational Intelligence, vol. 451, pp. 349–359. Springer, Berlin, Heidelberg (2013)
13. Allen, S., Burke, E.K., Hyde, M., Kendall, G.: Evolving reusable 3d packing heuristics with genetic programming. In: Proceedings of the 11th Annual conference on Genetic and evolutionary computation. pp. 931–938. GECCO'09, ACM, New York (2009)
14. Fukunaga, A.S.: Evolving local search heuristics for sat using genetic programming. In: Genetic and Evolutionary Computation—GECCO 2004, Lecture Notes in Computer Science, vol. 3103, pp. 483–494. Springer Berlin, Heidelberg (2004)
15. Hyde, M.R., Burke, E.K., Kendall, G.: Automated code generation by local search. J. Oper. Res. Soc. **64**(12), 1725–1741 (2013)

16. Hyde, M.: A Genetic programming hyper-heuristic approach to automated packing. Ph.D. thesis, University of Nottingham (2010)

17. Johnson, D.S., Demers, A., Ullman, J.D., Garey, M.R., Graham, R.L.: Worst-case performance bounds for simple one-dimensional packing algorithms. SIAM J. Comput. **3** (4), 299–325 (1974)

18. Yao, A.C.C.: New algorithms for bin packing. J. ACM **27**, 207–227 (1980)

19. Rhee, W.T., Talagrand, M.: On line bin packing with items of random size. Math. Oper. Res. **18**(2), 438–445 (1993)

20. Coffman, E., Jr., Galambos, G., Martello, S., Vigo, D.: Bin Packing Approximation Algorithms: Combinatorial Analysis. Kluwer Academic Publishers (1998)

21. Kämpke, T.: Simulated annealing: Use of a new tool in bin packing. Ann. Oper. Res. **16**, 327–332 (1988)

22. Falkenauer, E.: A hybrid grouping genetic algorithm for bin packing. J. Heuristics **2**, 5–30 (1996)

23. Ponce-Pérez, A., Pérez-Garcia, A., Ayala-Ramirez, V.: Bin-packing using genetic algorithms. In: Proceedings of the 15th International Conference on Electronics, Communications and Computers (CONIELECOMP 2005). pp. 311–314. IEEE Computer Society, Los Alamitos, CA, USA (2005)

24. Schwerin, P., Wäscher, G.: The bin-packing problem: A problem generator and some numerical experiments with ffd packing and mtp. Int. Trans. Oper. Res. **4**(5–6), 377–389 (1997)

25. O'Neill, M., Brabazon, A.: Measuring instance difficulty for combinatorial optimization problems. Comput. Oper. Res. **39**(5), 875–889 (2012)

26. Sotelo-Figueroa, M., Puga Soberanes, H., Martin Carpio, J., Fraire Huacuja, H., Cruz Reyes, L., Soria-Alcaraz, J.: Evolving and reusing bin packing heuristic through grammatical differential evolution. In: Nature and Biologically Inspired Computing (NaBIC), 2013 World Congress on. pp. 92–98 (2013)

27. Derrac, J., García, S., Molina, S., Herrera, F.: A practical tutorial on the use of nonparametric statistical tests as a methodology for comparing evolutionary and swarm intelligence algorithms. Swarm and Evolutionary Computation, pp. 3–18 (2011)

28. Garey, M.R., Johnson, D.S.: "Strong" np-completeness results: motivation, examples, and implications. J. ACM **25**, 499–508 (1978)

29. Martello, S., Toth, P.: Knapsack Problems Algorithms and Computer Implementations. Wiley, New York (1990)

30. Schoenfield, J.E.: Fast, exact solution of open bin packing problems without linear programming. Ph.D. thesis, US Army Space and Missile Defense Command, Huntsville, Alabama (2002)

31. Belov, G., Scheithauer, G.: A cutting plane algorithm for the one-dimensional cutting stock problem with multiple stock lengths. Eur. J. Oper. Res. **141**, 274–294 (2002)

32. Beasley, J.: Or-library: distributing test problems by electronic mail. J. Oper. Res. Soc. **41**(11), 1069–1072 (1990)

33. Scholl, A., Klein, R., Jürgens, C.: Bison: A fast hybrid procedure for exactly solving the one-dimensional bin packing problem. Comput. Oper. Res. **24**(7), 627–645 (1997)

34. Alvim, A., Ribeiro, C., Glover, F., Aloise, D.: A hybrid improvement heuristic for the one-dimensional bin packing problem. J. Heuristics **10**(2), 205–229 (2004)

35. Falkenauer, E., Delchambre, A.: A genetic algorithm for bin packing and line balancing. In: Proceedings of IEEE International Conference on Robotics and Automation, vol. 2, pp. 1186–1192 May 1992

36. Coffman, Jr., E.G., Garey, M.R., Johnson, D.S.: Approximation Algorithms for Bin Packing: A Survey. In: Hochbaum, D.S. (eds.) Approximation Algorithms for NP-hard Problems, pp. 46–93. PWS Publishing Co., Boston (1997)

37. Falkenauer, E.: Tapping the full power of genetic algorithm through suitable representation and local optimization: application to bin packing. In: Biethahn, J., Nissen, V. (eds.) Evolutionary Algorithms in Management Applications, pp. 167–182. Springer, Berlin (1995)

38. Gent, I.: Heuristic solution of open bin packing problems. J. Heuristics 3(1), 299 304 (1998)
39. Koza, J.R., Poli, R.: Genetic programming. In: Burke, E.K., Kendall, G. (eds.) Search Methodologies: Introductory Tutorials in Optimization and Decision Support Techniques, pp. 127–164. Kluwer, Boston (2005)
40. lan Fang, H., lan Fang, H., Ross, P., Ross, P., Corne, D., Corne, D.: A promising genetic algorithm approach to job-shop scheduling, rescheduling, and open-shop scheduling problems. In: Proceedings of the Fifth International Conference on Genetic Algorithms. pp. 375–382. Morgan Kaufmann (1993)
41. Sheskin, D.J.: Handbook of Parametric and Nonparametric Statistical Procedures. CRC, 2nd. edn. (2000)

Comparative Study of Particle Swarm Optimization Variants in Complex Mathematics Functions

Juan Carlos Vazquez, Fevrier Valdez and Patricia Melin

Abstract Particle Swarm Optimization (PSO) is one of the evolutionary computation techniques based on the social behaviors of birds flocking or fish schooling, biologically inspired computational search and optimization method. Since first introduced by Kennedy and Eberhart (A new optimizer using particle swarm theory 39–43, 1995 [1]) in 1995, several variants of the original PSO have been developed to improve speed of convergence, improve the quality of solutions found, avoid getting trapped in the local optima and so on. This paper is focused on performing a comparison of different approaches of inertia weight such as *constant, random adjustments, linear decreasing, nonlinear decreasing and fuzzy particle swarm optimization;* we are using a set of 4 mathematical functions to validate our approach. These functions are widely used in this field of study.

1 Introduction

Particle Swarm Optimization (PSO) was developed by Kennedy and Eberhart in 1995 [1, 2], and is a stochastic search method based on population. The idea behind this algorithm was inspired by the social behavior of animals, such as bird flocking or fish schooling. The process of the PSO algorithm in finding optimal values follows the work of this animal society. In a PSO system, a swarm of individuals (called particles) fly through the search space. Each particle represents a candidate solution to the optimization problem. The performance of each particle is measured using a fitness function that varies depending on the optimization problem. The PSO has been applied successfully to a number of problems, including standard function optimization problems [3–6], solving permutation problems [7, 8] and training multi-layer neural networks [5, 9, 10]. The basic PSO has problems with consistently converging to good solutions, so, there are several modifications that

J.C. Vazquez · F. Valdez (✉) · P. Melin
Tijuana Institute of Technology, Tijuana, Mexico
e-mail: fevrier@tectijuana.mx

© Springer International Publishing Switzerland 2015
O. Castillo and P. Melin (eds.), *Fuzzy Logic Augmentation of Nature-Inspired Optimization Metaheuristics*, Studies in Computational Intelligence 574,
DOI 10.1007/978-3-319-10960-2_11

163

have been proposed from the original PSO. It modifies to accelerate achieving of the best conditions, improving convergence of the PSO and increasing the diversity of the swarm. A brief description of some inertia weight approaches is presented below. A fuzzy particle swarm optimization is developed to improve the performance of PSO and is compared with basic modifications of PSO. Benchmark functions were used to measure the performance of the PSO algorithm with the different approaches. This paper is organized as follows: Sect. 2 describes the standard PSO algorithms, Sect. 3 describe the different variants of inertia weight used in this paper, Sect. 4 provides a brief description of fuzzy logic and describe the fuzzy system used in this paper, Sect. 5 presents simulation results for functions mathematic and finally, conclusions are summarized in Sects. 6.

2 Standard PSO Algorithm

The basic equations are usually given as follow:

$$v_{ij}(t+1) = vi_{ij}(t) + c_1 r_{1j}(t)[y_{ij}(t) - x_{ij}(t)] + c_2 r_{2j}(t)[\hat{y}_j(t) - x_{ij}(t)] \tag{1}$$

$$x_i(t+1) = x_{ij}(t) + v_{ij}(t+1) \tag{2}$$

The social network employed by the PSO reflects the star topology. For the star neighborhood topology, the social component of the particle velocity update reflects information obtained from all the particles in the swarm, referred to as $\hat{y}(t)$, where $v_i(t)$ is the velocity of particle i in dimension $j = 1, \ldots, n_x$ at time step t, $x_{ij}(t)$ is the position of particle i in dimension j at time step t, c_1 and c_2 are positive acceleration constants used to scale the contribution of the cognitive and social components respectively, and $r_{1j}(t)$, $r_{1j}(t) \sim U$ (0, 1) are random values in the range [0, 1], sampled from a uniform distribution. These random values introduce a stochastic element to the algorithm.

Let $x_i(t)$ denote the position of particle i in the search space at time step t, which denotes discrete time steps. The position of the particle is changed by adding a velocity, $v_i(t)$ to the current position. It is the velocity vector that drives the optimization process, and reflects both the experiential knowledge of the particle and socially exchanged information from the particle's neighborhood.

3 Variants of PSO

Several variants of the PSO algorithm have been developed [11–13]. It has been shown that the question of convergence of the PSO algorithm is implicitly guaranteed if the parameters are adequately selected [14, 15].

3.1 Inertia Weight

This variation was introduced by Shi and Eberthart [13] as a mechanism to control the exploration and exploitation abilities of the swarm, and as a mechanism to eliminate the need for velocity clamping [16]. The inertia weight was successful in addressing the first objective, but could not completely eliminate the need for velocity clamping. The inertia weight, w, basically works by controlling how much memory of the previous flight direction will influence the new velocity. The velocity equation is changed from Eq. (1) to:

$$v_{ij}(t+1) = wv_{ij}(t) + c_1r_{1j}(t)[y_{ij}(t) - x_{ij}(t)] + c_2r_{2j}(t)[\hat{y}_j(t) - x_{ij}(t)] \qquad (3)$$

The value of w is extremely important to ensure convergent behavior, and to optimally tradeoff exploration and exploitation. For $w \geq 1$, velocities increase over time, accelerating towards the maximum velocity (assuming velocity clamping is used), and the swarm diverges. For $w < 1$, particles decelerate until their velocities reach zero (depending on the values of the deceleration coefficients). Large values for w facilitate exploration, with increased diversity. Very small values eliminate the exploration ability of the swarm. Little momentum is then preserved from the previous time step, which enables quick changes in direction. The smaller w, the more do the cognitive and social components control position updates.

As with the maximum velocity, the optimal values for the inertia weight is problem dependent [15]. Initial implementations of the inertia weight used a static value for the entire search duration, for all particles for each dimension. Later implementations made use of dynamically changing inertia values. These approaches usually start with large inertia values, which decrease over time to smaller values. In doing so, particles are allowed to explore in the initial search steps, while favoring exploitations as time increases. The choice of values for w has to be made in conjunction with the selection of the values for c_1 and c_2.

Approaches to dynamically varying the inertia weight can be grouped into the following categories:

- **Random adjustments,** where a different inertia is randomly selected at each iteration. One approach is to sample from a Gaussian distribution, e.g.

$$w \sim N(0.72, \sigma) \qquad (4)$$

where σ is small enough to ensure that w is not predominantly greater than one. Alternatively, Peng et al. used [17]

$$w = (c_1r_1 + c_2r_2) \qquad (5)$$

with no random scaling of the cognitive and social components.

- **Linear decreasing,** where an initially large inertia weight (usually 0.9) is linearly decreased to a small value (usually 0.4). From Naka et al. [18], Yoshida et al. [19]

$$w(t) = (w(0) - w(n_t)) \frac{(n_t - t)}{n_t} + w(n_t) \qquad (6)$$

Where n_t is the maximum number of time steps for which the algorithm is executed, $w(0)$ is the initial inertia weight, $w(n_t)$ is the final inertia weight, and $w(t)$ is the inertia at time step t. Note that $w(0) > w(n_t)$.

- **Nonlinear decreasing,** where an initially large value decreases nonlinearly to a small value. Nonlinear decreasing methods allow a shorter exploration time than the linear decreasing methods, with more time spent on refining solutions (exploiting). Nonlinear decreasing methods will be more appropriate for smoother search space. The following nonlinear methods have been defined:

  - From Naka et al. [18],

$$w(t + 1) = \frac{(w(t) - 0.4)(n_t - t)}{n_t + 0.4} \qquad (7)$$

  With $w(0) = 0.9$.

- **Increasing inertia,** where the inertia is linearly increased from 0.4 to 0.9 [18, 20].

The linear and nonlinear adaptive inertia methods above are very similar to the temperature schedule of simulated annealing [21].

4 Fuzzy Logic

Fuzzy logic was investigated for the first time in the mid-sixties at the University of California Berkeley by the brilliant Iranian engineer Zadeh [22], when he realized what it is called the principle of incompatibility: "As the complexity of a system increases, our ability to build precise and instructions on their behavior decreases to the threshold beyond which the accuracy and meaning are mutually exclusive characteristic." This principle is the basic to the fuzzy systems used today. The structure of a Fuzzy System consists of the following elements:

1. The rule base that contains linguistic rules provided by the expert, or can be extracted from numerical data.
2. The fuzzifier, which assigns to the numerical entries in their corresponding membership function. This is needed to activate rules, which are specified in term of linguistic variables; fuzzifier takes the input values and determines the degree to which they belong to each of the fuzzy sets via membership functions.
3. The fuzzy inference engine defines the allocation of input fuzzy sets to the corresponding values of the output fuzzy sets. This determines the degree to which each part of the antecedent is satisfied for each rules.

4. The defuzzifier in the case of Mamdani-assigned sets of outputs in a number that is a fact not fuzzy. Given a fuzzy set which manages a range of output values, the defuzzifier returns a number within a set of fuzzy numbers.

4.1 Fuzzy System

A fuzzy particle swarm optimization is developed to improve the performance of PSO; where the inertia weight w and learning factories c_1 and c_2 are dynamically adjusted on the basic of fuzzy sets and rules during the evolution process [23, 24]. We define a fuzzy system for parameter adaptation to consist of the following components:

- Two inputs, one to represent the number of generation for unchanged best fitness (NU), and the other the current value of the inertia weight (W).
- Three output, learning (c1 and c2) and the change in inertia weight (chw).
- Three fuzzy sets, namely LOW, MEDIUM and HIGH, respectively implemented as a left triangle, triangle and right triangle membership function.
- Nine fuzzy rules from which the change in inertia is calculated. An example rule in the fuzzy is:
  If normalized NU is LOW, and

    W value is LOW
  then the C1 is LOW, C2 is LOW and chW is HIGH

The range of NU is normalized into [0, 1.0], the value for w is bounded in $0.2 \leq w \leq 1.2$ and the values of c_1 and c_2 are bounded in $1.0 \leq c_1, c_2$ 2.0.

5 Simulation Results for Mathematical Functions

In this section four mathematical functions are tested under the different reviewed approaches of inertia weight.

5.1 F1 Function

$$|x| + \cos(x) \tag{12}$$

where $-100 < x_n < 100$ $n = 1$, n is the number of the variable to be optimized. The objective is to find the minimum of F1 function and the related variable locations.

Fig. 1 Surface plot of the F1
function in n = 1 variable

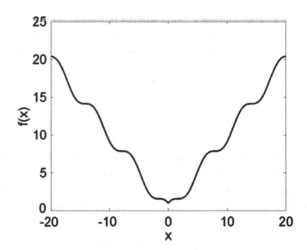

Figure 1 shows the surface plot of the F1 function in one variable. There is a unique minimum point in the figure with location $f(0) = 1$.

For the simulations we used a population size of 20, 40, 60, 80 and 100. $c_1 = c_2 = 2$, Constant inertia weight (W) = 0.8, 100,000 generations and 8 experiments.

Table 1 shows the optimization results of F1 function, where W is the inertia weight with different approaches (Constant, Random adjustments, Linear decreasing, Nonlinear decreasing), *Tswarm* is the population size, *Gen* is average of generations, *Success rate* is the reached minimum in the generation *Gen* and the number of variables is 1 (Table 2).

5.2 F2 Function

$$x \sin(4x) + 1.1y \sin(2y) \tag{13}$$

where $0 < x_n < 10$ $n = 2$, n is the number of variables to be optimized. The objective is to find the minimum of F2 function and the related variable locations. Figure 2 shows the surface plot of the F2 function in one variable. There is a unique minimum point in the figure with location $f(0.9039, 0.8668) = -18.5547$.

For the simulations we used a population size of 20, 40, 60, 80 and 100. $c_1 = c_2 = 2$, Constant inertia weight (W) = 0.8, 100,000 generations and 8 experiments.

Table 3 shows the optimization results of F2 function, where W is the inertia weight with different approaches (Constant, Random adjustments, Linear decreasing, Nonlinear decreasing), *Tswarm* is the population size, *Gen* is average of

Table 1 Optimization results of F1 function with the different approaches of W

Number of variables	W	Constant		Random adjustments		Linear decreasing		Nonlinear decreasing	
	Tswar	Gen	Success rate	Gen	Success rate	Gen	Success rate	Gen	Success rate
1	20	482	1	163.4	1	12944	1	167.3	1
1	40	451	1	157.5	1	13,780.4	1	153.6	1
1	60	420.2	1	141.7	1	13,252.8	1	146.8	1
1	80	433.6	1	140.3	1	12,362.4	1	80.4	1
1	100	427	1	137.4	1	12,790.6	1	45.4	1

Table 2 Optimization results of F1 function with adjust of fuzzy parameter (W, c_1, c_2) with 1variable

Number of variables	Tswar	Gen	Success rate
1	40	263.7	1
1	60	246.1	1
1	80	246.9	1
1	100	240.9	1
1	200	234.9	1
1	300	226.9	1
1	400	228.2	1

Fig. 2 Surface plot of the F2 function in $n = 2$ variables

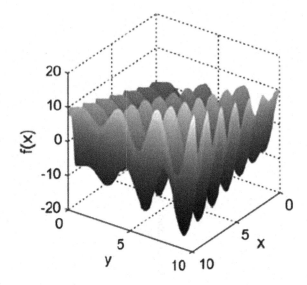

generations, *Success rate* is the reached minimum in the generation *Gen* and the number of variables is 2 (Table 4).

5.3 F3 Function

$$f(x) = \sum_{n=1}^{N-1} \left\{ 100 \left[x_{n+1} - x_n^2 \right]^2 + \left[1 - x_n \right]^2 \right\}$$ (14)

where $-100 < x_i < 100$, $i = 1, 2,..., $ n, n is the number of variables to be optimized. The objective is to find the minimum of F3 function and the related variable locations. Figure 3 shows the surface plot of the F3 function in one variable. There is a unique minimum point in the figure with location $f(1, 1) = 0$.

Table 3 Optimization results of F2 function with the different approaches of W

Number of variables	W	Constant	Success rate	Random adjustments	Success rate	Linear decreasing	Success rate	Nonlinear decreasing	Success rate
	Tswar	Gen		Gen		Gen		Gen	
2	20	98.2	−18.5547	44.1	−18.5547	1630.2	−18.5547	51	−18.5547
2	40	76.4	−18.5547	28.6	−18.5547	1037	−18.5547	24.6	−18.5547
2	60	66.4	−18.5547	29.4	−18.5547	867	−18.5547	18.8	−18.5547
2	80	71.8	−18.5547	27.8	−18.5547	380.9	−18.5547	53.3	−18.5547
2	100	58.2	−18.5547	23.5	−18.5547	279.4	−18.5547	63.6	−18.5547

Table 4 Optimization results of F2 function with adjust of fuzzy parameter (W, c_1, c_2) with 2 variables

Number of variables	Tswar	Gen	Success rate
2	40	41.7	−18.5547
2	60	34.7	−18.5547
2	80	33.9	−18.5547
2	100	33.2	−18.5547
2	200	26.7	−18.5547
2	300	24.8	−18.5547
2	400	22	−18.5547

Fig. 3 Surface plot of the F3 function in n = 2 variables

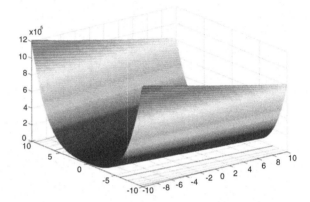

For the simulations we used a population size of 20, 40, 60, 80 and 100. $c_1 = c_2 = 2$, Constant inertia weight $(W) = 0.8$, 100,000 generations and 8 experiments.

Table 5 shows the optimization results of F3 function, where W is the inertia weight with different approaches (Constant, Random adjustments, Linear decreasing, Nonlinear decreasing), *Tswarm* is the population size, *Gen* is average of generations, *Success rate* is the reached minimum in the generation *Gen* and the number of variables is 8 (Table 6).

5.4 F4 Function

$$10N + \sum_{n=1}^{N} \left[x_n^2 - 10\cos(2\pi x_n) \right] \tag{15}$$

where $-100 < x_i < 100$, $i = 1, 2,\ldots,$ n, n is the number of variables to be optimized. The objective is to find the minimum of F2 function and the related variable locations. Figure 2 shows the surface plot of the F2 function in one variable. There is a unique minimum point in the figure with location $f(0,0) = 0$.

Table 5 Optimization results of F3 function with the different approaches of W

Number of variables	W	Constant	Success rate	Random adjustments	Success rate	Linear decreasing	Success rate	Nonlinear decreasing	Success rate
	Tswar	Gen		Gen		Gen		Gen	
8	20	100,000	44.2	100,000	125,145.9	100,000	75.6	100,000	250,047.1
8	40	100,000	1.9	100,000	1,273.5	100,000	32.6	100,000	126,306.5
8	60	100,000	44.2	100,000	290.3	100,000	54.1	100,000	125,167
8	80	100,000	0.2,101	100,000	32.4	100,000	53.2	100,000	11.8
8	100	100,000	0.7,971	100,000	0.0168	100,000	43.2	100,000	10.8

Table 6 Optimization results of F3 function with adjust of fuzzy parameter (W, c_1, c_2) with 8 and 10 variables

Number of variables	Tswar	Gen	Success rate	Number of variables	Tswar	Gen	Success rate
8	60	100,000	3.0252e-26	10	60	100,000	2.159e-10
8	80	100,000	8.659 e-27	10	80	100,000	1.533e-25
8	100	100,000	2.168e-26	10	100	100,000	1.535e-23
8	200	100,000	1.373e-27	10	200	100,000	4.283e-24
8	300	100,000	2.885e-28	10	300	100,000	5.167e-22
8	400	100,000	1.498e-28	10	400	100,000	1.015e-26

Fig. 4 Surface plot of the F4 function in n = 2 variables

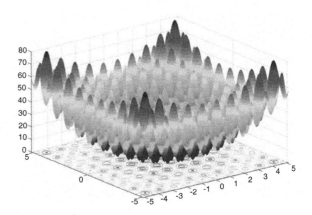

For the simulations we used a population size of 20, 40, 60, 80 and 100. $c_1 = c_2 = 2$, Constant inertia weight (W) = 0.8,10,000 generations and 8 experiments (Fig. 4).

Table 7 shows the optimization results of F4 function, where W is the inertia weight with different approaches (Constant, Random adjustments, Linear decreasing, Nonlinear decreasing), *Tswarm* is the population size, *Gen* is average of generations, *Success rate* is the reached minimum in the generation *Gen* and the number of variables is 8 (Tables 8 and 9).

Table 7 Optimization results of F4 function with the different approaches of W

Number of variable s	W	Constant	Success rate	Random adjustments	Success rate	Linear decreasing	Success rate	Nonlinear decreasing	Success rate
	Tswar	Gen		Gen		Gen		Gen	
8	20	100,000	2.1,889	100,000	2.6117	33,227.8	0	100,000	5.5966
8	40	100000	0.3979	75560.4	1.4924	32305.3	0	100000	4.6016
8	60	100000	0.3979	50465	0.4948	31269.1	0	100000	2.2386
8	80	6534	0	27040.1	0.2487	30979.4	0	62773.5	1.2437
8	100	2151.8	0	39308.5	0.4974	30566.4	0	63317.3	0.7462

Table 8 Optimization results of F4 function with adjust of fuzzy parameter (W, c_1, c_2) with 8 and 10 variables

Number of variables	Tswar	Gen	Success rate	Number of variables	Tswar	Gen	Success rate
8	40	3740.1	0	10	40	4018.5	0
8	60	2023	0	10	60	2761.3	0
8	80	1852.3	0	10	80	1955.4	0
8	100	1471.9	0	10	100	1935.6	0
8	200	836.6	0	10	200	1236.4	0
8	300	701.9	0	10	300	877.2	0
8	400	533.6	0	10	400	1145.6	0

Table 9 Optimization results of F4 function with adjust of fuzzy parameter (W, c_1, c_2) with 20 and 30 variables

Number of variables	Tswar	Gen	Success rate	Number of variables	Tswar	Gen	Success rate
20	40	38206	0	30	40	39357.7	0
20	60	18821	0	30	60	28866.2	0
20	80	10017	0	30	80	22006	0
20	100	5983.7	0	30	100	17492	0
20	200	3297.2	0	30	200	8994	0
20	300	3783.9	0	30	300	9641.5	0
20	400	2824.1	0	30	400	7856	0

6 Conclusions

In this paper, we proposed a comparative study of inertia weight approaches. Simulation tests on four mathematical functions were performed to compare the proposed study. The simulation results of the different approaches of inertia weight show that among more variables has the function the PSO used more generations to find the minimum, therefore results show great difference in both the constant inertia weight as its approaches. Was applied a fuzzy particle swarm optimization showing better difference between the results of the different approaches of inertia weight, noteworthy that the adjustment was both the value of inertia weight (w) as acceleration coefficients (c_1 and c_2), therefore, we increased the number of variables to see better the performance of fuzzy PSO, showing significant results. Parameters can be adjusted of fuzzy PSO (input variables, type of membership and rules) to improve results as future work.

References

1. Eberhart, R.C., Kennedy, J.: A new optimizer using particle swarm theory. In: Proceedings of the Sixth International Symposium on MicroMachine and Human Science, pp. 39–43 (1995)
2. Kennedy, J., Eberhart, R.C.: Particle swarm optimization. In: Proceedings of the IEEE International Joint Conference on Neuronal Networks, pp. 1942–1948. IEEE Press (1995)
3. Angeline, P.J.: Evolutionary optimization versus particle swarm optimization: philosophy and performance differences. In: Proceeding of the Seventh Annual Conference on Evolutionary Programming, pp. 601–610 (1998)
4. Kennedy, J., Spears, W.: Matching algorithms to problems: an experimental test of the particle swarm and some genetic algorithms on the multimodal problem generator. In: Proceedings of the IEEE Congress on Evolutionary Computation, pp. 78–83. IEEE Press (1998)
5. Valdez, F., Melin, P., Castillo, O.: Evolutionary method combining particle swarm optimization and genetic algorithms using fuzzy logic for decision making. In: Proceedings of the IEEE International Conference on Fuzzy Systems, pp. 2114–2119 (2009)
6. Valdez, Fevrier, Melin, Patricia, Castillo, Oscar: An improved evolutionary method with fuzzy logic for combining particle swarm optimization and genetic algorithms. Appl. Soft Comput. **11**(2), 2625–2632 (2011)
7. Salerno, J.: Using the particle swarm optimization technique to train a recurrent neural model. In: Proceedings of the IEEE International Conference on Tools with Artificial Intelligence, pp 45–49. IEEE Press (1997)
8. Tasgetiren, M.F., Liang, Y.C., Sevkli, M., Gencyilmaz, G.: A particle swarm optimization algorithm for makespan and total flowtime minimization in the permutation flowshop sequencing problem. Eur. J. Oper. Res. **177**, 1930–1947 (2007)
9. Ribeiro, P.F., Kyle Schlansker, W.: A Hybrid Particle Swarm and Neuronal Network Approach for Reactive Power Control. IEEE (2006)
10. Russell, C., Eberthart, Hu X.: Human Tremor Analysis Using Particle Swarm Optimization, Purdue School of Engineering and Technology. Indiana University Purdue University Indianapolis, Indianapolis (1999)
11. Shi, Y.H., Eberhart, R.C.: Fuzzy adaptive particle swarm optimization. In: IEEE International Conference on Evolutionary Computation, pp. 101–106 (2001)
12. Kennedy, J.: The behaviour of particles. Evol. Progr. **VII**, 581–587 (1998)
13. Shi, Y.H., Eberhart, R.C.: A modified particle swarm optimizer. In Proceedings of the IEEE International Conference on Evolutionary Computation, pp. 69–73 (1998)
14. Cristian, T.I.: The particle swarm optimization algorithm: convergence analysis and parameter selection. Inf. Process. Lett. **85**(6), 317–325 (2003)
15. Shi, Y.H., Eberhart, R.C.: Parameter selection in particle swarm optimization. In: Porto, V.W., Waagen, D. (eds.) EP 1998. LNCS, vol. 1447, pp. 591–600. Springer, Heidelberg (1998)
16. Eberhart, R.C., Shi, Y.: Particle swarm optimization: developments, applications and resources. In: Proceedings of the IEEE Congress on Evolutionary Computation, vol. 1, pp. 27–30. IEEE Press (2001)
17. Peng, J., Chen, Y., Eberhart, R.C.: Battery Pack State of charge estimator design using computational intelligence approaches. In Proceedings of the Annual Battery Conference on Applications and Advances, pp. 173–177 (2000)
18. Naka, S., Genji, T., Yura, T., Fukuyama, Y.: Practical distribution state estimation using hybrid particle swarm optimization. In: IEEE Power Engineering Society Winter Meeting, vol. 2, pp. 815–820 (2001)
19. Yoshida, H., Fukuyama, Y., Takayuma, S., Nakanishi, Y.: A particle swarm optimization for reactive power and voltage control in electric power systems considering voltage security assessment. In: Proceedings of the IEEE International Conference on Systems, Man, and Cybernetics, vol. 6, pp. 497–502 (1999)

20. Zheng, Y., Ma, L., Zhang, L., Qian, J.: Empirical study of particle swarm optimizer with increasing inertia weight. In: Proceedings of the IEEE Congress on Evolutionary Computation, pp. 221–226, IEEE Press (2003)

21. Kirkpatrick, S., Gelatt, C.D., Vecchi, M.P.: Optimization by simulated annealing. Science **220**, 671–680 (1983)

22. Zadeh, L.: Fuzzy logic. IEEE Comput. **1**, 83 (1988)

23. Zhang, W., Liu, Y.: Multi-objective reactive power and voltage control based on fuzzy optimization strategy and fuzzy adaptive particle swarm. Electrical Power and Energy Systems (2008)

24. Shi, Y., Eberhart, R.C.: Fuzzy adaptive particle swarm optimization. In: Proceedings of the IEEE Congress on Evolutionary Computation, vol. 1, pp. 101–106. IEEE Press (2001)

25. Brits, R., Engelbrecht, A.P., van den Bergh, F.: A Niching particle swarm optimizer. In: Proceedings of the Fourth Asia-Pacific Conference on Simulated Evolution and Learning, pp. 692–696 (2002)

26. Carlisle, A., Dozier, G.: Adapting particle swarm optimization to dynamic environments. PhD thesis, Auburn University (2002)

27. Clerc, M.: The swarm and the queen: towards a deterministic and adaptive particle swarm optimization. In: Proceedings of the IEEE Congress on Evolutionary Computation, vol. 3, pp. 1951–1957 (1999)

28. Clerc, M., Kennedy, J.: The particle swarm-explosion, stability, and convergence in a multidimensional complex space. IEEE Trans. Evol. Comput. **6**(1), 58–73 (2002)

29. Eberhart, R.C., Shi, Y.: Comparing inertia weights and constriction factors in particle swarm optimization. In: Proceedings of the IEEE Congress on Evolutionary Computation, vol. 1, pp. 84–88 (2000)

Optimization of Modular Network Architectures with a New Evolutionary Method Using a Fuzzy Combination of Particle Swarm Optimization and Genetic Algorithms

Fevrier Valdez

Abstract We describe in this paper a new hybrid approach for optimization combining Particle Swarm Optimization (PSO) and Genetic Algorithms (GAs) using Fuzzy Logic to integrate the results. The new evolutionary method combines the advantages of PSO and GA to give us an improved FPSO + FGA hybrid method. Fuzzy Logic is used to combine the results of the PSO and GA in the best way possible. Also fuzzy logic is used to adjust parameters in the FPSO and FGA.

1 Introduction

We describe in this paper a new evolutionary method combining PSO and GA, to give us an improved FPSO + FGA hybrid method. We apply the hybrid method to optimize the architectures of Modular Neural Networks (MNNs) for pattern recognition. We also apply the hybrid method to mathematical function optimization to validate the new approach. In this case, we are using the Rastrigin's function, Rosenbrock's function, Ackley's function, Sphere's function Griewank's function, Michalewics's function, Zakharov's function, Dixon's function, Levy's function and Perm's function [1, 2, 3].

The paper is organized as follows: in part 2 a description about the genetic algorithms for optimization problems is given, in part 3 the Particle Swarm Optimization is presented, the FPSO + FGA method is presented in part 4 and 5, the simulation results for modular neural network optimization are presented in part 6, finally we can see the conclusions reached after the study of the proposed evolutionary computing method.

F. Valdez (✉)
Tijuana Institute of Technology, Tijuana, B.C, Mexico
e-mail: fevrier@tectijuana.mx

© Springer International Publishing Switzerland 2015
O. Castillo and P. Melin (eds.), *Fuzzy Logic Augmentation of Nature-Inspired Optimization Metaheuristics*, Studies in Computational Intelligence 574,
DOI 10.1007/978-3-319-10960-2_12

179

2 Genetic Algorithm for Optimization

John Holland, from the University of Michigan initiated his work on genetic algorithms at the beginning of the 1960s. His first achievement was the publication of *Adaptation in Natural and Artificial System* [4] in 1975.

He had two goals in mind: to improve the understanding of natural adaptation process, and to design artificial systems having properties similar to natural systems [5].

The basic idea is as follows: the genetic pool of a given population potentially contains the solution, or a better solution, to a given adaptive problem. This solution is not "active" because the genetic combination on which it relies is split between several subjects. Only the association of different genomes can lead to the solution.

Holland's method is especially effective because it not only considers the role of mutation, but it also uses genetic recombination, (crossover) [6]. The crossover of partial solutions greatly improves the capability of the algorithm to approach, and eventually find, the optimal solution.

The essence of the GA in both theoretical and practical domains has been well demonstrated [7]. The concept of applying a GA to solve engineering problems is feasible and sound. However, despite the distinct advantages of a GA for solving complicated, constrained and multi-objective functions where other techniques may have failed, the full power of the GA in application is yet to be exploited [8, 9].

In Fig. 1 we show the reproduction cycle of the Genetic Algorithm.

The Simple Genetic Algorithm can be expressed in pseudo code with the following cycle:

1. Generate the initial population of individuals aleatorily P(0)
2. While (number _ generations <= maximum _ numbers _ generations)

Do:

    {

        Evaluation;
        Selection;
        Reproduction;
        Generation ++;

Fig. 1 The reproduction cycle

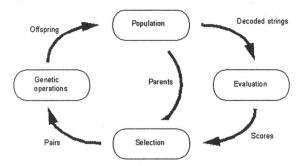

}

3. Show results
4. End

3 Particle Swarm Optimization

Particle swarm optimization (PSO) is a population based stochastic optimization technique developed by Eberhart and Kennedy in 1995, inspired by social behavior of bird flocking or fish schooling [10].

PSO shares many similarities with evolutionary computation techniques such as Genetic Algorithms (GA) [11]. The system is initialized with a population of random solutions and searches for optima by updating generations. However, unlike the GA, the PSO has no evolution operators such as crossover and mutation. In the PSO, the potential solutions, called particles, fly through the problem space by following the current optimum particles [12].

Each particle keeps track of its coordinates in the problem space, which are associated with the best solution (fitness) it has achieved so far (The fitness value is also stored). This value is called *pbest*. Another "best" value that is tracked by the particle swarm optimizer is the best value, obtained so far by any particle in the neighbors of the particle. This location is called *lbest*. When a particle takes all the population as its topological neighbors, the best value is a global best and is called *gbest* [13].

The particle swarm optimization concept consists of, at each time step, changing the velocity of (accelerating) each particle toward its p*best* and *lbest* locations (local version of PSO). Acceleration is weighted by a random term, with separate random numbers being generated for acceleration toward *pbest* and *lbest* locations [13].

In the past several years, PSO has been successfully applied in many research and application areas. It is demonstrated that PSO gets better results in a faster, cheaper way compared with other methods [14, 15].

Another reason that PSO is attractive is that there are few parameters to adjust. One version, with slight variations, works well in a wide variety of applications. Particle swarm optimization has been used for approaches that can be used across a wide range of applications, as well as for specific applications focused on a specific requirement [14, 16-18].

The pseudo code of the PSO is as follows
For each particle

Initialize particle

End
Do

For each particle

    Calculate fitness value
    If the fitness value is better than the best fitness value (pBest) in history

        set current value as the new pBest

End
Choose the particle with the best fitness value of all the particles as the gBest
For each particle

    Calculate particle velocity
    Update particle position

End

While maximum iterations or minimum error criteria is not attained

4 FPSO + FGA Method

This method combines the characteristics of PSO and GA using several fuzzy systems for integration of results and parameter adaptation. In this section, the proposed FPSO + FGA method is presented.

The general idea of the proposed FPSO + FGA method can be seen in Fig. 2. The method can be described as follows:

1. A mathematical function to be optimized is received.
2. The role of both FPSO and FGA is evaluated.
3. A main fuzzy system is responsible for receiving values resulting from step 2.
4. The main fuzzy system decides which method to use (FPSO or FGA)
5. Another fuzzy system receives the values of Error and DError as inputs to evaluate if it is necessary to change the parameters in FPSO or FGA.
6. There are 3 fuzzy systems. One is for decision making (is called 'fuzzymain'), the second one is to change the parameters of the GA (is called 'fuzzyga'), in this case change the value of the crossover and mutation rate and the third fuzzy system is used to change the parameters of the PSO (is called 'fuzzypso') in this case change the values of the cognitive acceleration 'c1', and social acceleration 'c2'. The main fuzzy system (called 'fuzzymain') decides in the final step the optimum value for the function introduced in step 1.
7. Repeat the above steps until the termination criterion of the algorithm is met.

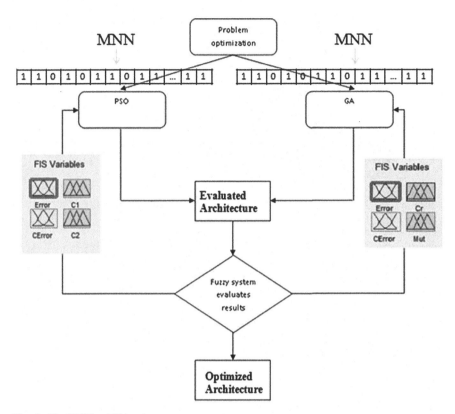

Fig. 2 The FPSO + FGA scheme

5 Full Model of FPSO + FGA

The basic idea of the FPSO + FGA scheme is to combine the advantage of the individual methods using a fuzzy system for decision making and the others two fuzzy systems to improve the parameters of the FGA and FPSO when is necessary (Fig. 3).

As can be seen in the proposed hybrid FPSO + FGA method, it is the internal fuzzy system structure, which has the primary function of receiving as inputs (Error and DError) the results of the FGA and FPSO outputs. The fuzzy system is responsible for integrating and decides which are the best results being generated at run time of the FPSO + FGA. It is also responsible for selecting and sending the problem to the "fuzzypso" fuzzy system when the FPSO is activated or to the "fuzzyga" fuzzy system when FGA is activated. Also activating or temporarily stopping depending on the results being generated. Figure 4 shows the membership functions of the main fuzzy system that is implemented in this method. The fuzzy system is of Mamdani type because it is more common in this type of fuzzy control and the defuzzification method is the centroid. In this case, we are using this type of

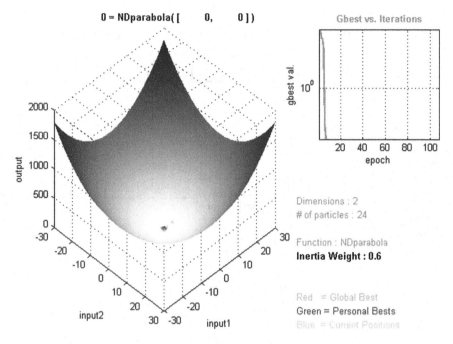

Fig. 3 Simulation of sphere's function with FPSO

defuzzification because in other papers we have achieved good results [19]. The membership functions are of triangular form in the inputs and outputs as is shown in Fig. 4. Also, the membership functions were chosen of triangular form based on past experiences in this type of fuzzy control. The fuzzy system consists of 9 rules. For example, one rule is if error is P and DError is P then best value is P (view Fig. 5). Figure 6 shows the fuzzy system rule viewer. Figure 7 shows the surface corresponding to this fuzzy system. The other two fuzzy systems are similar to the main fuzzy system.

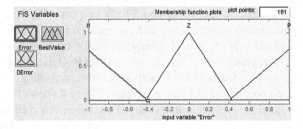

Fig. 4 Fuzzy system membership functions

Fig. 5 Fuzzy system rules

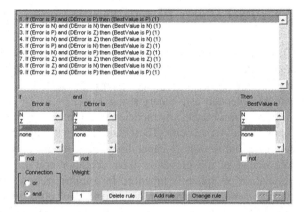

Fig. 6 Fuzzy system rules viewer

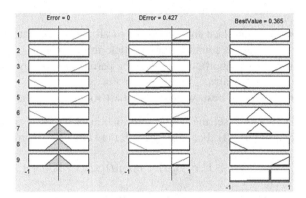

Fig. 7 Surface of fuzzy system

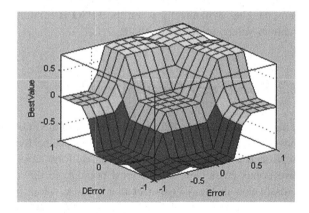

5.1 FPSO (Fuzzy Particle Swarm Optimization)

This section presents a detailed description of the FPSO model. The classical representation scheme for GAs is binary vectors of fixed length. In the case of an n_x—dimensional search space, each individual consists of n_x variables with each variable encoded as a binary string.

The swarm is typically modeled by particles in multidimensional space that have a position and a velocity. These particles fly through hyperspace (i.e., R^n) and have two essential reasoning capabilities: their memory of their own best position and knowledge of the global or their neighborhood's best. In Fig. 3 we can see a simulation of Sphere's function. In a minimization optimization problem, "best" simply meaning the position with the smallest objective value. Members of a swarm communicate good positions to each other and adjust their own position and velocity based on these good positions. So a particle has the following information to make a suitable change in its position and velocity:

- A global best that is known to all and immediately updated when a new best position is found by any particle in the swarm.
- The neighborhood best that the particle obtains by communicating with a subset of the swarm.
- The local best, which is the best solution that the particle has seen.

In this case, the social information is the best position found by the swarm, referred as $\hat{y}(t)$. For gbest FPSO, the velocity of particle i is calculated as

$$v_{ij}(t+1) = vi_j(t) + c_1 r_{1j}(t)[y_{ij}(t) - x_{ij}(t)] + c_2 r_{2j}(t)[\hat{y}_j(t) - x_{ij}(t)] \qquad (1)$$

where $vi_j(t)$ is the velocity of particle i in dimension $j = 1, \ldots, n_x$ at time step t, $xi_j(t)$ is the position of particle i in dimension j at time step t, C_1 and C_2 represents the cognitive and social acceleration. In this case, these values are fuzzy because they are changing dynamically when the FPSO is running, and $r_{1j}(t)$, $r_{2j} \sim U(0, 1)$ are random values in the range [0, 1].

5.2 FGA (Fuzzy Genetic Algorithm)

This section presents a detailed description of the FGA. Several crossover operators have been developed for GAs, depending on the format in which individuals are represented. For binary representations, uniform crossover, one point crossover and two points cross over are the most popular. In this case we are using two points crossover with fuzzy crossover rate because we are adding a fuzzy system called 'fuzzyga' that is able of change the crossover and mutation rate.

5.3 Definition of the Fuzzy Systems Used in FPSO + FGA

'fuzzypso': In this case we are using a fuzzy system called 'fuzzypso', and the structure of this fuzzy system is as follow:

Number of Inputs: 2

Number of Outputs: 2

Number of membership functions: 3

Type of the membership functions: Triangular

Number of rules: 9

Defuzzification: Centroid

The main function of the fuzzy system called 'fuzzypso' is to adjust the parameters of the PSO. In this case, we are adjusting the following parameters: 'c1' and 'c2'; where:

'c1' = Cognitive Acceleration

'c2' = Social Acceleration

We are changing these parameters to test the proposed method. In this case, with 'fuzzypso' is possible to adjust in real time the 2 parameters that belong to the PSO.

'fuzzyga': In this case we are using a fuzzy system called **'fuzzyga'**, the structure of this fuzzy system is as follows:

Number of Inputs: 2

Number of Outputs: 2

Number of membership functions: 3

Type of membership functions: Triangular

Number of rules: 9

Defuzzification: Centroid

The main function of the fuzzy system called 'fuzzypso' is to adjust the parameters of the GA. In this case, we are adjusting the following parameters: 'mu', 'cr'; where:

'mu' = mutation

'cr' = crossover

'fuzzymain': In this case, we are using a fuzzy system called **'fuzzymain'**. The structure of this fuzzy system is as follows:

Number of Inputs: 2

Number of Outputs: 1

Number of membership functions: 3

Type of membership functions: Triangular

Number of rules: 9

Defuzzification: Centroid

The main function of the fuzzy system, called 'fuzzymain' is to decide on the best way for solving the problem, in other words if it is more reliable to use the FPSO or FGA. This fuzzy system is able to receive two inputs, called error and derror, it is to evaluate the results that are generated by FPSO and FGA in the last step of the algorithm.

6 Simulation Results for Modular Neural Network for Optimization

Several tests of the FPSO + FGA method for MNN optimization were made in the Matlab programming language.

All the implementations were developed using a computer with quad core2 processor of 64 bits that works to a frequency of clock of 2.5 GHz, 4 GB of RAM Memory and Windows Vista operating system.

We describe below simulation results of our approach for face recognition with modular neural networks (MNNs). We used two-layer feed-forward MNNs with the Conjugated-Gradient training algorithm [19]. The challenge is to find the optimal architecture of this type of neural network, which means finding out the optimal number of layers and nodes of the neural network [20]. We are using the Yale face database [16] that contains 165 grayscale images in GIF format of 15 individuals, for this paper only 10 subjects were used for training the MNN. There are 5 images per subject, one per different facial expression: center-light, happy, left-light, normal and right-light.

In total 50 images were used (see Fig. 9). Three images per subject were used for training the MNN and the other two for the recognition. Regarding the genetic algorithm for NN evolution, we used a hierarchical chromosome for representing the relevant information of the network. First, we have the bits for representing the number of layers of the MNN, in this case, the initial topology was of 3 modules with 2 layers per module with 500 neurons in the first layer, 300 neurons in the second layer in each module. Therefore we used a representation the 2415 bits in total (view Fig. 8). The PSO is organized in a similar fashion, but there is less number of parameters. In Fig. 11 we can see the architecture of a MNN that we are using with the evolutionary proposed method FPSO + FGA.

The fitness function used in this case for the MNN combines the information of the error objective and also the information about the number of nodes as a second objective. This is shown in the following equation.

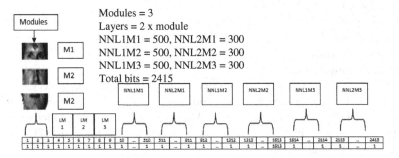

Fig. 8 Binary representation for FPSO + FGA (No optimized)

Fig. 9 Images of the Yale
face database

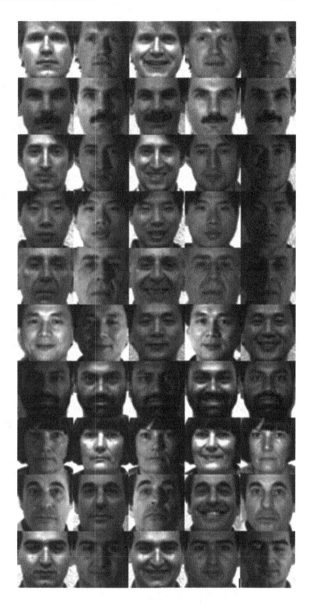

$$f(z) = \left(\frac{1}{\alpha * Ranking(ObjV1) + \beta * ObjV2} \right) * 10 \qquad (2)$$

The first objective is basically the average sum of squared of errors as calculated
by the predicted outputs of the MNN compared with real values of the function.
This is given by the following equation.

$$f_1 = \frac{1}{N} \sum_{i=1}^{N} (Y_i - y_i)^2 \tag{3}$$

The second objective is the complexity of the neural network, which is measured by the total number of nodes in the architecture.

The final topology of the neural network for the problem of face recognition is obtained by the hybrid evolutionary method FPSO + FGA. The comparison of the final objective values (errors) will be shown in the following section. In the final architecture, the result of the MNN evolution is a particular architecture with different number of nodes by layers. Several tests were made; we obtained different optimized architectures for this Modular Neural Network; the best architecture obtained was the following:

Layers = 2 x module
NNL1M1 = 90, NNL2M1 = 50
NNL1M2 = 100, NNL2M2 = 150
NNL1M3 = 70, NNL2M3 = 90
Total bits = 565
Where:
NNL1M1 = Number of neurons of the layer 1 in module 1.
NNL1M1 = Number of neurons of the layer 1 in module 1.
NNL2M1 = Number of neurons of the layer 2 in module 1.
NNL1M2 = Number of neurons of the layer 1 in module 2.
NNL2M2 = Number of neurons of the layer 2 in module 2.
NNL1M3 = Number of neurons of the layer 1 in module 3.
NNL2M3 = Number of neurons of the layer 2 in module 3.

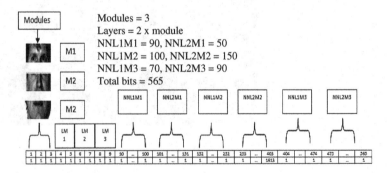

Fig. 10 Binary representation optimized for FPSO + FGA

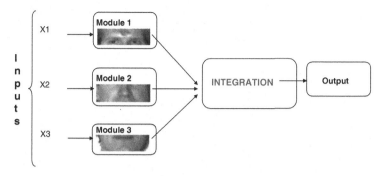

Fig. 11 Architecture of the modular neural network

We can see in the Fig. 10 the binary representation for this optimized architecture. With this final topology the Neural Network was trained and the ten images
were recognized. It can be seen in table I the different architectures obtained with
this method. The proposed method optimizes the initial architecture proposed for
the problem of face recognition (Fig. 11).

The Parameters of Table 1 and 2, are as follows:

Ima = Number of images. Mod = Number of modules. NNL = Number of
neurons Layer 1, 2 or 3. GE = Goal Error. RE = Reached Error.
IDENT = Number of images recognized. Train M = Training Method for the
MNN. FGA = Parameters of Fuzzy Genetic Algorithm. FPSO = Parameters of
Fuzzy Genetic Algorithm. VAR = Number of variables for the mathematical
function.

Also this FPSO + FGA have been applied, for optimization of complex mathematical functions to validate our approach. The Table 2, shows the simulation
results with 10 mathematical functions. It can be seen in Table 2 that this method is
good alternative to solve this type of problems. The mathematical functions are
evaluated with 2, 4, 8 and 16 variables. The mean was calculated after running the
FPSO + FGA 50 times. The parameters in FPSO + FGA as crossover, mutation,
social and cognitive acceleration are fuzzy, because are changing dynamically when
the method is running, this is main characteristic of this method to find the best
results.

Table 1 Simulation results for the MNN

IMA	Mod	LMod	NN L1M1	NN L2M1	NN L1M2	NNL 2M2	NNL 1M3	NNL2M3	GE	RE	IDENT	TR M
10	3	2	20	60	80	50	60	120	0.01	0.03	8	CGrad
10	3	2	90	50	100	150	70	90	0.01	0.005	10	CGrad
10	3	2	70	40	80	40	90	30	0.01	0.02	8	CGrad
10	3	2	150	135	200	90	84	40	0.01	0.003	9	CGrad
10	3	2	100	120	100	145	100	70	0.01	0.001	10	CGrad

Table 2 Simulation results for Mathematical Functions with FPSO + FGA

Math function	Variables = 2		Variables = 4		Variables = 8		Variables = 16	
Math	BEST	MEAN	BEST	MEAN	BEST	MEAN	BEST	MEAN
Rastrigin	1.45E-06	3.05E-04	0.00034	0.0755	0.01318	0.9311	0.548397150613653	5.0946
Rosenbrock	1.17E-02	1.17E-02	0.0285	0.5991	0.15800	3.8925	0.25555	4.33334
Ackley	8.42E-04	4.98E-03	8.42E-01	4.98E-02	0.7	1.56	2.35	2.63
Sphere	5.75E-11	1.05E-10	1.946e-05	4.5109e-004	0.00059	0.0057	0.00248	0.0211
Griewank	7.88E-11	1.07E-07	7.18130761856450e-06	1.1182e-004	0.000162543802633697	9.2993e-004	0.000408945493635127	0.0043
Michalewics	−1.8010	−1.8201	−1.80129624818704	−1.8002	−1.80130169498191	−1.8005	−1.80130138163081	−1.8004
Zakharov	6.00E-07	0.00168	3.23737992547765e-07	8.4129e-005	1.33084082865516e-07	6.4901e-005	8.63410514832145e-07	7.0065e-005
Dixon	0.0070	0.007	0.0036	0.1343	0.0581	0.7676	0.94444	4.5968
Levy	0.000001	0.00026	0.000001	0.0031	0.0058	0.00229	0.0111	0.1093
Perm	0.0001	0.0068	0.1488	5.6553	0.0018	7.3555	0.0075	8.5670

7 Conclusions

The analysis of the simulation results of the evolutionary method considered in this paper, FPSO + FGA lead us to the conclusion that for the optimization of Modular Neural Networks with this method is a good alternative because it is easier to optimize the architecture of Modular Neural Network than to try it with PSO or GA separately. This is, because the combination PSO and GA with fuzzy rules gives a new hybrid method FPSO + FGA. It can be seen in Table 1 that the second and five architectures obtained after applying FPSO + FGA recognize the ten images, and as a consequence we are demonstrating that it is reliable for this type of applications. Recently we are working with more images to test the effectiveness of this approach. Also, we can appreciate that this method has been tested with 10 benchmark mathematical functions to validate our approach. In Table 2 we can see been the simulation results obtained with the method.

Acknowledgments We would like to express our gratitude to CONACYT, Universidad Autónoma de Baja California and Tijuana Institute of Technology for the facilities and resources granted for the development of this research.

Reference

1. Valdez, F., Melin P.: Parallel Evolutionary Computing using a cluster for Mathematical Function Optimization, pp. 598–602. Nafips, San Diego (2007)
2. Montiel, O., Castillo, O., Melin, P., Rodriguez, A., Sepulveda, R.: Human evolutionary model: a new approach to optimization. Inf. Sci. **177**(10), 2075–2098 (2007)
3. Valdez, F., Melin, P., Castillo, O.: Evolutionary computing for the optimization of mathematical functions. In: Analysis and Design of Intelligent Systems Using Soft Computing Techniques. Advances in Soft Computing, vol. 41. Springer, Berlin (2007)
4. Holland, J.H.: Adaptation in Natural and Artificial System. The University of Michigan Press, Ann Arbor (1975)
5. Goldberg, D.: Genetic Algorithms. Addison Wesley, Boston (1988)
6. Emmeche, C.: Garden in the Machine. The Emerging Science of Artificial Life, p. 114. Princeton University Press, Princeton (1994)
7. Man, K.F., Tang, K.S., Kwong S.: Genetic Algorithms: Concepts and Designs. Springer, Berlin (1999)
8. Back, T., Fogel, D.B., Michalewicz, Z. (eds.): Handbook of Evolutionary Computation. Oxford University Press, Oxford (1997)
9. Castillo, O., Valdez F., Melin P.: Hierarchical genetic algorithms for topology optimization in fuzzy control systems. Int. J. General Syst. **36**(5), 575–591 (2007)
10. Kennedy, J., Eberhart, R.C.: Particle swarm optimization. In: Proceedings of IEEE International Conference on Neural Networks, Piscataway, NJ, pp. 1942–1948 (1995)
11. Fogel, D.B.: An introduction to simulated evolutionary optimization. IEEE Trans. Neural Netw. **5**(1), 3–14 (1994)
12. Angeline, P.J.: Using selection to improve particle swarm optimization. In: Proceedings 1998 IEEE World Congress on Computational Intelligence, pp. 84–89. Anchorage, Alaska, IEEE (1998)

13. Kennedy, J., Mendes, R.: Population structure and particle swarm performance. In: Proceeding of IEEE Conference on Evolutionary Computation, pp. 1671–1676 (2002)

14. Angeline, P.J.: Evolutionary Optimization Versus Particle Swarm Optimization: Philosophy and Performance Differences, Evolutionary Programming VII, Lecture Notes in Computer Science 1447, pp. 601–610. Springer, Berlin (1998)

15. Kennedy, J., Mendes.: The particle swarm-explosion, stability, and convergence in a multidimensional complex space. IEEE Trans. Evol. Comput. **6**(1), 58–73 (2002)

16. Wei, W., Jiatao, S., Zhongzxiu, Y., Zheru.: Wavelet-based illumination compensation for face recognition using Eifenface method intelligent control and automation. WCICA **2**, 10356–10360 (2006)

17. Eberhart, R.C., Kennedy, J.: A new optimizer using particle swarm theory. In: Proceedings of the Sixth International Symposium on Micromachine and Human Science, Nagoya, Japan, pp. 39–43 (1995) (J.-G. Lu, Title of paper with only the first word capitalized. *J. Name Stand. Abbrev.*, in press)

18. Veeramachaneni, K., Osadciw, L., Yan, W.: Improving Classifier Fusion Using Particle Swarm Optimization. In: IEEE Fusion Conference. Italy (2006)

19. Castillo, O., Huesca, G., Valdez, F.: Proceedings of the International Conference on Artificial" Intelligence, IC-AI '04, Las Vegas, Nevada, USA, vol. 1, pp. 98–104, 21–24 June 2004

20. Castillo, O., Melin, P.: Hybrid intelligent systems for time series prediction using neural networks, fuzzy logic, and fractal theory. IEEE Trans. Neural Netw. **13**(6), 1395–1408 (2002)

Printed in the United States
By Bookmasters